Starfleet Guides
Volume II

GALILEO-CLASS STARCRUISER

Ship's Organization and Regulations Manual

RESTRICTED DISTRIBUTION
For Permanent Ship's Personnel Only

©1995 by *PARSEC Publications*, a division of
PAUMER ASSOCIATES INTERNATIONAL, INC.
P.O. Box 44395
Rio Rancho, NM 87174–4395

Written by: W. Paul Hollingsworth
Illustrated by: Evangeline "Eve" Burge

ISBN: 1–886810–01–X

First Printing: February 1995
Second Printing: February 1997

STARFLEET COMMAND
Office of the Chief of Operations
San Francisco, Earth

A MESSAGE FROM THE CHIEF OF OPERATIONS,
STARFLEET COMMAND

Congratulations on your assignment to a GALILEO-class Starcruiser. Designed specifically for scientific research and exploration of the furthest reaches of Federation space and beyond, this ship is equipped with state-of-the-art equipment and laboratories, and manned by some of the most qualified and experienced crewmembers in Starfleet.

You have been specifically selected as a crew member because of your demonstrated excellence in your particular field. Only the best of the best have even been considered for duty aboard a GALILEO-class Starcruiser. You have met that test. Your Captain and your crewmates expect no less of you now that you have reported aboard.

Life aboard your ship will present you with operational procedures and personal circumstances which may be new to you. In some cases they may appear to be exotic and confusing. Careful reading of this Ship's Organization and Regulations Manual will introduce you to your new surroundings and provide answers to most of your questions.

Should you encounter difficulties or have questions not addressed by this document, contact your Division Officer for assistance.

Welcome aboard one of Starfleet's finest ships. May you find your tour of duty aboard your new ship a fascinating journey into the unknown.

Kitryk Donaree Brannon
Admiral, Starfleet Command

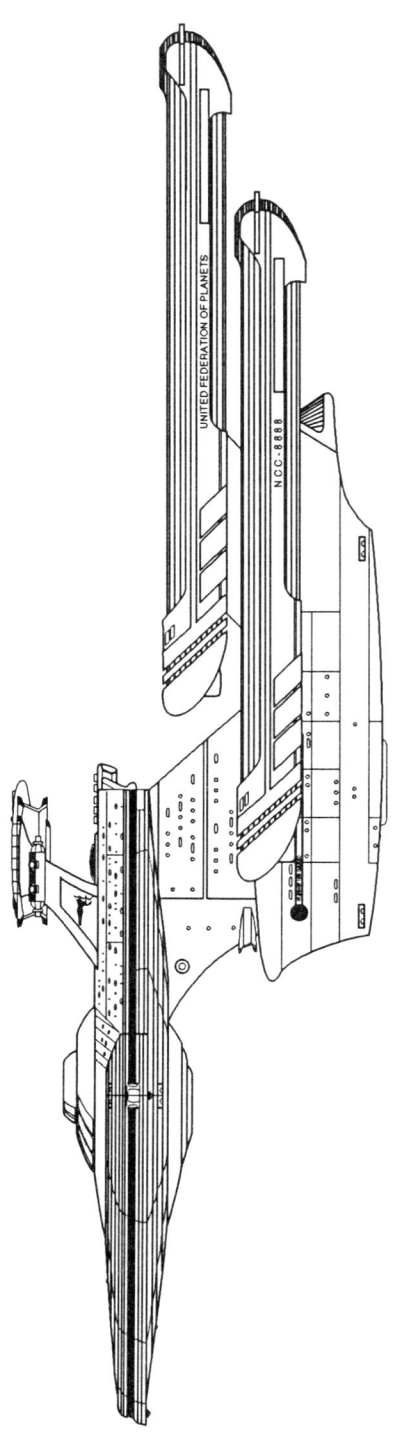

TABLE OF CONTENTS

CLASS INFORMATION

1 HISTORY

The famous Camp Khitomer peace conference of 2293 drastically altered the relationship between the United Federation of Planets and the Klingon Empire. Although the conference did not completely eradicate the atmosphere of mistrust and suspicion which had existed between the two governments since 2218, it did have a profound impact on Starfleet.

With the constant threat of imminent hostilities reduced to cautious co-existence, a major paradigm shift began—away from starships as primarily weapon platforms. The original concept of Starfleet in the Articles of Federation (Article 54) was as an armed peace-keeping force and a major instrument of scientific exploration and investigation. It was this latter aspect which now came under close scrutiny from the Federation Council.

Starfleet was directed by the Military Staff Committee to develop appropriate plans to expand its role in exploration and scientific study and relegate its peace-keeping duties to a secondary priority. This directive resulted in two studies. The first, from Starfleet, addressed the implementation of a more peaceful role for its starships. The second, from a group of militant ex-Starfleet officers and members of the defense industry complex, addressed the role of Starfleet in its entirety.

Starfleet laid out a revolutionary plan to convert older starships to unarmed exploration vessels which would then be transferred to the Bureau of Exploration and Research. This newly conceived organization of the United Federation of Planets would then assume all responsibilities for Starfleet functions unrelated to peace keeping.

While this concept would temporarily reduce Starfleet's starship inventory, the monies involved in maintaining and upgrading the obsolete vessels would immediately be reapportioned for construction of new classes of war ships. Almost as an afterthought, a recommendation was included to design and build a new class of exploration and scientific research ships with no offensive capabilities.

The second report contradicted Starfleet's plan at almost every turn. Its basic precept was that relations with the Klingon Empire were destined to deteriorate into active hostilities again. Concerns were also raised over the Romulan Star Empire and the impact of the Romulan/Klingon agreement on the security of Federation worlds. These situations, the report argued, made the decision to reduce the size of Starfleet, along with the recommendation to expend scare resources on building unarmed exploration vessels, a serious threat to Federation security and, in fact, bordered on treason.

Opposing viewpoints on the matter resulted in a schism not only in the Federation Council, but in Starfleet as well. The ensuing conflict of ideologies resulted in Council member resignations, withdrawal from the Federation by several governments, and the early retirement of several high-ranking Starfleet officers. A compromise, not entirely satisfactory to either side, was reached in late 2294.

Starfleet would continue to support both the peace-keeping and exploration roles—no new organization would be created—with exploration and scientific study established as a primary mission. Peace-keeping duties would remain, of course, but as a secondary responsibility. Appropriations were made to design and build a new class of starships, the Starcruiser, to be the vanguard of this new policy. In order to conserve resources, the Starcruiser would also be armed for peace-keeping duties.

Several hulls in the Starfleet inventory were considered for modification. Among these were the GRISSOM-class Science Vessel, the AVENGER-class Heavy Frigate, the ENTERPRISE II-class Heavy Cruiser, and the BELKNAP-class Strike Cruiser. In each case, the high cost of design changes required to accept the state-of-the-art technology envisioned for the Starcruiser forced the Corps of Engineers to abandon the idea of modifying an existing platform.

Finally, the Department of Starship Design, Development and Testing, Starfleet Corps of Engineers, decided that it would be more cost-effective to design the ship from the keel up. After 20 months of intensive effort, the CONSTELLATION-class Starcruiser was unveiled. A definite departure from established starship designs, the Starcruiser featured a laminated hull—a wider and thicker addition sandwiched between the upper and lower primary hull surfaces. Perhaps the most striking innovation was a double-tandem, perpendicular nacelle design which would provide maximum redundancy in case of warp engine damage. AT 325,000 metric tons displacement, it would be the largest vessel in Starfleet.

With a great deal of fanfare, USS CONSTELLATION (NCC-1017) was launched in 2299 and turned over to Starfleet's Cathedral Unit for shake down. The last message received from the CONSTELLATION indicated a matter/antimatter imbalance had created a wormhole into which the ship was being drawn.

This disaster spelled the end of the CONSTELLATION-class Starcruiser. Throughout its development and construction, anti-Starcruiser forces had repeatedly charged that the placement of the warp engine nacelles would cause matter/antimatter/impulse engine interference and result in damage or destruction of the ship. When their dire predictions came true, the design was scrapped and an effort begun to replace it.

The ten largest shipbuilding entities in the Federation were supplied with specifications and requested to prepare new designs for a ship to take the CONSTELLATION's place. Two years later, Sandia Shipbuilding and Conversion, Albuquerque Division, delivered a Starcruiser design which, with minimal changes, became the GALILEO-class Starcruiser.

A composite vessel, it married the best features of the AVENGER-class Heavy Frigate, the ENTERPRISE II-class Heavy Cruiser, and the BELKNAP-class Strike Cruiser. At 305,000 metric tons, it would still be the largest ship in Starfleet, but the decrease in tonnage from the original CONSTELLATION-class design proved to increase the GALILEO-class' performance significantly.

2 MISSION

The basic mission concept for the Starcruiser was defined in the Design Specification Package as "a vessel capable of:

(1) operating independently of other starships and support facilities for extended periods of time;

(2) providing extensive scientific laboratories and appropriate personnel to effect intensive and comprehensive investigations of geological, hydrological, biological, meteorological, and astronomical events and anomalies; and,

(3) playing a pivotal role as a major weapons platform and a Fleet-action coordination vessel in any conceivable offensive or defensive scenario."

Although the primary mission of a vessel is normally listed first in the Design Specification Package, the requirement for the Starcruiser to operate independently for extended periods of time is the cornerstone upon which the remaining two ship's capabilities depend. To meet this requirement, a number of changes to normal internal starship design and equipment had to be made. Six major areas were considered critical: environment, ergonomics, replication facilities, stores, power plant, and weapons systems. For a more detailed discussion of these areas, see the appropriate sections elsewhere.

Scientific investigation is the primary mission of the Starcruiser. Virtually the entire extended hull is dedicated to science laboratories. The capabilities of these state-of-the-art facilities are equalled only by the largest planetary science complexes.

As a Fleet-Action Coordination vessel, the Starcruiser becomes the flag ship for a Task Force Commander, who controls assigned assets (vessels, starbases, and/or planetary facilities) from the ship. Depending upon the military or political situation, the Task Force assumes responsibility for any one of a number of missions, including interdiction, blockade, defense, or offense. The Task Force mission can change rapidly as circumstances evolve.

The ship's normal complement is augmented by a number of strategic and tactical specialists. They man the Situation Room and provide command, control, and communications support to the Task Force Commander (for additional information on the Situation Room, see Page 99). A MIMAD (see Page 135) may be brought aboard the ship if deemed necessary.

3 SHIPS OF THE CLASS

A total of 12 GALILEO-class Starcruisers were constructed by Sandia Shipbuilding and Conversion at their Earth Orbit Spacedock Facilities. Each is named after a famous scientist or explorer in Federation history.

Name	Hull Number
USS GALILEO GALILEI	NCC-8888
USS ALBERT EINSTEIN	NCC-8889
USS CAPTAIN JAMES COOK	NCC-8890
USS CHANDRASEKHAR	NCC-8891
USS FERDINAND MAGELLAN	NCC-8892
USS JACQUES-IVES COUSTEAU	NCC-8893
USS LEONARDO DA VINCI	NCC-8894
USS ZEFRAM COCHRANE	NCC-8895
USS NICHOLAUS COPERNICUS	NCC-8896
USS CAROLUS LINNAEUS	NCC-8897
USS PAUL A. M. DIRAC	NCC-8898
USS STEPHEN W. HAWKING	NCC-8899

Starcruiser Names and Hull Numbers

3.01 CONSTRUCTION SCHEDULE

The original construction schedule called for up to five Starcruisers to be built simultaneous at the Earth Spacedock Facility of Sandia Shipbuilding and Conversion. Total construction time from keel laying to the commissioning ceremony for each vessel was to be five Earth years. An early-completion clause in the contract awarded additional credits to the builder for every day cut from each ship's delivery date.

Five MAYA-class drydocks (*Kirtland Burke*, *Summer Lianne daVinci*, *Meredith Beaton*, *Valerian Lynn*, and *Judith V. Hill*) were available at Sandia's Earth Orbit Facilities for use by the construction team . Five others were transferred from the Mars Orbit Facility as current projects were completed (*Crystal Singer*, *Gaylynne Robinson*, *Sandra Vaglienti*, *Kasey Standish*, and *Mary Kay Skopinski*). Two years into the construction cycle, two additional drydocks were built (*Rena Pacini* and *Carolyn Carmines*). A total of 12 drydocks were in use during the last four years of the contract.

Through this concentration of 85% of Sandia Shipbuilding and Conversion's total assets, and the use of Pre-positioning-Just-in-Time component procurement concepts, Sandia decreased total construction time for each Starcruiser to approximately four years and four months.

Hull Number	Keel Laid	Christened	Commissioned
NCC-8888	5 OCT 2302	20 JUL 2304	15 FEB 2307
NCC-8889	15 JAN 2303	10 OCT 2304	9 MAY 2307
NCC-8890	5 APR 2303	20 JAN 2305	30 JUL 2307
NCC-8891	28 JUL 2303	13 MAY 2305	23 NOV 2307
NCC-8892	17 OCT 2303	1 AUG 2305	11 FEB 2308
NCC-8893	6 JAN 2304	21 OCT 2305	1 MAY 2308
NCC-8894	24 APR 2304	8 FEB 2306	18 AUG 2308
NCC-8895	16 JUL 2304	1 JUN 2306	11 NOV 2308
NCC-8896	20 OCT 2304	4 AUG 2306	2 FEB 2309
NCC-8897	16 JAN 2305	31 OCT 2306	5 MAY 2309
NCC-8898	8 APR 2305	1 JUL 2307	2 AUG 2309
NCC-8899	31 JUL 2305	15 FEB 2307	25 AUG 2309

Starcruiser Construction Timeline

This unprecedented concentration of corporate assets, coupled with the fact that every major Starfleet inspection was satisfactory, allowed the 12 Starcruiser to be ready for service years before the original schedule. The early delivery of two ships, USS CHANDRASEKHAR (NCC-8891) and USS STEPHEN W. HAWKING (NCC-8899), forced Starfleet to delay their transfer to active service due to crew shortages.

3.02 LOGOS AND MOTTOS

Each Star Cruise has its own logo and motto. Although the official use of individual ship logos ceased in 2277, the practice has continued in spite of that Starfleet decision. Before that time, logos were designed and assigned to ships by the Heraldry Division of Starfleet Corps of Engineers. Since 2277, the unofficial logos have been prepared by The Friends of Starfleet, an organization of retired Starfleet officers and civilians with acute interest in or close ties to Starfleet.

USS GALILEO GALILEI
NCC-8888

Starcruiser

Keel Laid—5 OCT 2302 Christened—20 JUL 2304 Commissioned—15 FEB 2307

**Sandia Shipbuilding and Conversion
Albuquerque Division
Earth Spacedock Facilities**

Prima Inter Pares
First Among Her Equals

Typical Commissioning Plaque

Most GALILEO-class Starcruisers affix their unofficial logo to each wing of the Weapons Bridge, just above its connection point with the primary hull.

This practice serves no tactical or strategic purpose, since ships can readily be identified by their sensor signature. Although not condoned, it is tacitly ignored by Starfleet.

A ship's motto is normally chosen by the Commissioning Crew (or Plank-owners, as they are traditionally termed). Contests are often held with the crewmember submitting the winning entry receiving a suitable reward the day the ship is commissioned. A commissioning plaque containing the ship's motto is installed on the Main Bridge.

Ship	Motto
NCC-8888	*Prima Inter Pares* (First Among Her Equals)
NCC-8889	The Universe Is Relative
NCC-8890	To Sail the Farthest Reaches
NCC-8891	To Search, Therefore to Find
NCC-8892	Above All, Knowledge
NCC-8893	To Know the Infinite Unknown
NCC-8894	Second Star on the Right, and On Till Morning
NCC-8895	We Can Because We Think We Can
NCC-8896	Courage Leads to the Stars
NCC-8897	A Tall Ship and a Star
NCC-8898	There Are Always Alternatives
NCC-8899	Today the Improbable, Tomorrow the Impossible

Starcruiser Mottos

USS GALILEO GALILEI

USS CAPTAIN JAMES COOK

USS ALBERT EINSTEIN

USS CHANDRASEKHAR

USS FERDINAND MAGELLAN

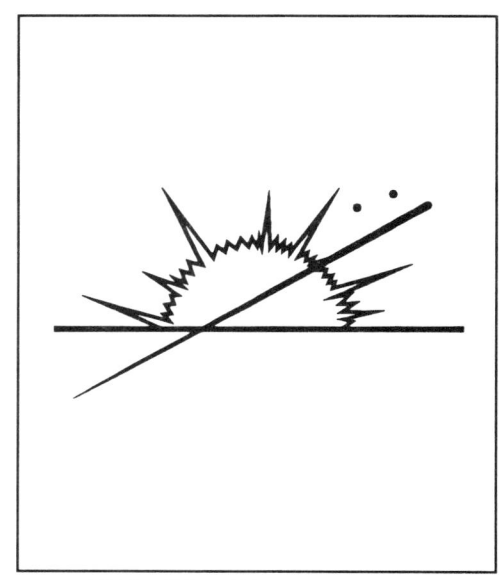

USS LEONARDO DAVINCI

USS JACQUES-IVES COUSTEAU

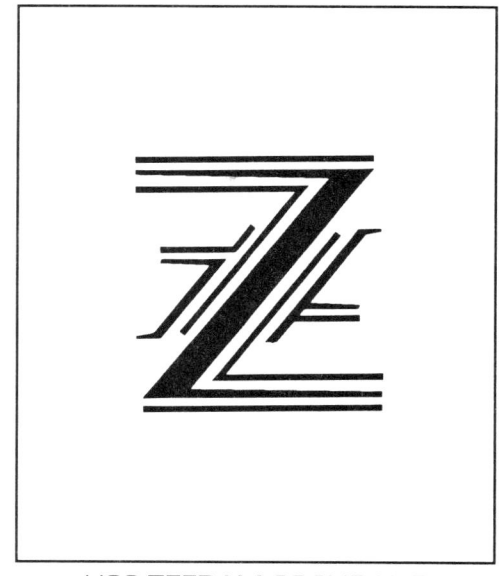

USS ZEFRAM COCHRANE

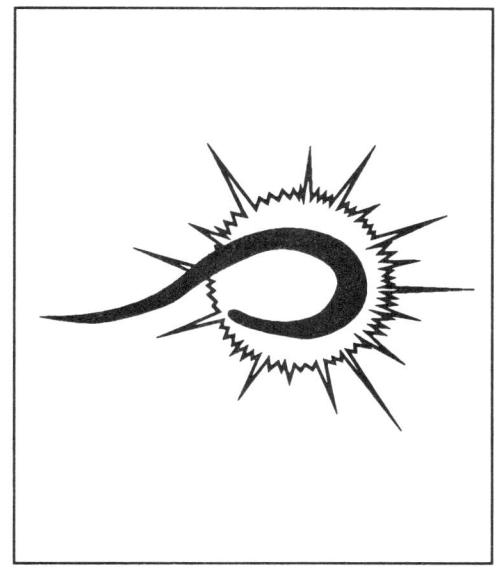

USS NICHOLAUS COPERNICUS

USS PAUL A. M. DIRAC

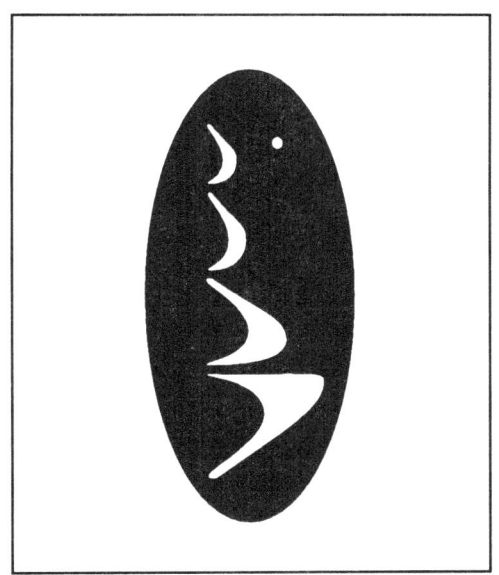

USS CAROLUS LINNAEUS

USS STEPHEN W. HAWKING

3.03 QUADRANT ASSIGNMENTS

Each Quadrant of the Federation Treaty Zone has three Starcruisers assigned to it. Normally, each ship operates independently, even though each is assigned to a specific division in the numbered fleet responsible for that particular Quadrant.

Should the Starcruiser be required to act as Fleet-Action Coordinators during Fleet exercises or in the event of a crisis, it is treated as a major asset for the Fleet Commander to use as strategy and tactics dictate.

Ship	Quadrant	Division
NCC-8888	3	31
NCC-8889	4	41
NCC-8890	3	32
NCC-8891	1	11
NCC-8892	1	12
NCC-8893	2	21
NCC-8894	4	42
NCC-8895	2	22
NCC-8896	1	13
NCC-8897	2	23
NCC-8898	4	43
NCC-8899	3	33

Starcruiser Quadrant Assignments

3.04 HOME PORTS

Each ship in Starfleet is assigned a home port, where the vessel is primarily based. For Starcruisers, however, this is more of a formality than a fact. The mission of a Starcruiser is seldom satisfied by remaining in close proximity to any one planet or starbase. Instead, the Starcruiser is often light years away from any support facilities. The ships do return to their assigned home port on an infrequent basis, but not as a routine occurrence.

Ship	Home Port	System
NCC-8888	Daran VI	Daran
NCC-8889	Democritus	Demos 372 Alpha
NCC-8890	Arcturus	Alpha Boötis
NCC-8891	Bardex III	Bardex
NCC-8892	Cochrane I	Zeta Riguli
NCC-8893	Barsoom	Delta Gamma
NCC-8894	Springboard	Alpha Canaris
NCC-8895	Ramillies II	Ramillies
NCC-8896	Dundee II	Dundee
NCC-8897	Sigma Draconis	Sigma Beta 433
NCC-8898	Hodlahr	Sigma Indus 462
NCC-8899	Raven	Gamma Lyrai

Starcruiser Homeport Assignments

4　CRITICAL COMPONENTS

A number of specific, technical requirements were included in the Design Specification Package for the GALILEO-class Starcruiser. Some could be met by utilizing systems, components, or concepts already in use by other Starfleet vessels. In some cases, however, off-the-shelf solutions were unsatisfactory, and sub-contracts were let to supply unique designs based on emerging technology.

The following section discusses various critical shipboard areas, systems, or concepts in the Starcruiser design.

4.01　ARMAMENT

Included in the weapons suite aboard GALILEO-class Starcruisers are the improved multi-ported, ECHO-enhanced phasers, the Mega-phasers first utilized by the AVENGER-class heavy frigate (also ECHO-enhanced), and photon torpedo launchers which are standard equipment on Starfleet vessels. This vast array of defensive/offensive armament establishes the Starcruiser as the single most powerful weapons platform available in Starfleet at this time. Its superior weapons systems were deemed essential to its mission of exploring space outside Federation boundaries and to its secondary role as a Fleet-Action Coordination vessel.

4.01.01　WEAPONS BRIDGE

Photon torpedo tubes and Mega-phaser emplacements are located in the Weapons Bridge. This arrangement was first utilized on the AVENGER-class Heavy Frigate and resulted in a major reduction in the space required for weapons systems. The Bridge curves above the extended portion of the primary hull with the photon torpedo tubes (fore-and-aft firing) in the center. Mega-phasers are located on the port and starboard wings.

Fifty photon torpedoes are maintained on automated launch conveyors within the Weapons Bridge, 25 in each wing. These torpedoes are ready at all times for immediate launch by the Weapons Control Officer or the Navigator (depending upon circumstances). As soon as one torpedo is launched, another is ready for transfer from the conveyor to the launch rail. At periodic intervals, every torpedo is moved to a Weapons Room for inspection and maintenance. As the torpedoes are transferred off the conveyor, other torpedoes are moved into the launch complex to replace them.

Triple-redundant power conduits supply the Mega-phasers with energy. Each of these conduits is heavily armored for protection. Each conduit is designed to handle (as a safeguard) a 25-percent overload from the Weapons Power Distribution System. Should all three conduits on one wing be damaged, auxiliary power can be routed from the conduits on the undamaged wing. However, these emergency supply conduits can provide only 50 percent of the power provided by the main supply conduits. This decreased supply reduces the maximum output and range of the Mega-phaser.

At the base of each wing, where the Weapons Bridge becomes an integral part of the hull, there is a Weapons Room. Maintenance on photon torpedoes is accomplished in this guarded facility. One-half of the remaining on-board torpedo inventory is also kept here. The balance of torpedoes are stored in an armored magazine located in the secondary hull.

4.01.02 FIRE CONTROL SYSTEMS

The primary Fire-Control Guidance System for torpedoes and Mega-phaser control is located in the torpedo bulge atop the Weapons Bridge. A special sensor package, used only by the fire control system, is an integral part of the torpedo bulge's exterior casing. This system is capable of operating in either a passive or active mode.

When operating in the passive mode, target emissions are used as a lock-on source for the selected weapon. This mode is used when the ship is in Emissions Control (EMCON) condition. Setting EMCON requires that all equipment which radiates a detectable signal (such as sensors, fire control systems, and transporters) be shut down. Range and accuracy of the fire control system in passive mode are severely compromised.

Full sensor input is fed into the fire control system in the active mode. Because targeting information in the active mode is reflected as well as radiated, range and accuracy are extremely high. Detection of the Star cruiser in this mode is simplified for an enemy vessel since the ship becomes a "hot spot" of emissions throughout the spectrum.

If used as a Fleet-Action Coordination vessel, the lateral sensor array and the Operations computer assume all duties of the fire control system. Inputs from all starships with the Starcruiser's control are fed into the Operations computer and a consolidated tactical picture is provided each ship in return. Depending upon the size of the Fleet-Action Task Force, the Starcruiser can provide a clear picture of both allied and enemy vessels or facilities within a volume of space up to 1,288.25 cubic light years.

Should battle damage disable the fire control sensor package, input from the OMNI-Synse lateral sensor array can be routed to the fire control guidance system. For a more detailed discussion of the Fire Control Computer itself, see Page 53.

4.01.03 MEGA-PHASERS

To augment the standard phaser banks, when more concentrated firepower is required, each Starcruiser is outfitted with ECHO-enhanced Mega-phasers. Two of these fore-and-aft firing units are located on the Weapons Bridge, one on the port side and one on the starboard.

Designed to project a pulsed harmonic beam onto a target, each produces a maximum output of 2.6×10^{12} Megajoules at a maximum range of 1,000,000 kilometers. These data assume a stationary target and a stationary weapons platform. If either, or both, are in motion, the amount of destructive energy delivered to the target becomes a function of their relative vectors and is therefore reduced.

The ECHO enhancement (Enhanced Collimation via Harmonic Oscillation) is effected by a δ-wave generator slaved to the fire-control computer. Based on the input parameters, the emitted beam is encased by δ waves adjusted to coincide with the harmonic spikes. This δ jacket protects the radiation from shield, gravimetric, and warp envelope dispersal effects, which can degrade the effective levels of Energy on Target.

δ radiation is lethal to all known life forms. ECHO enhancement is not to be used in any situation where preservation of life on the target is of concern.

4.01.04 PHASERS

The **PHAS**ed **E**nergy **R**ectification weapons system is used primarily to disable a target. Each of the 12 phaser banks on a Starcruiser is composed of two synchronized nozzle/chamber units. The nozzles are mounted on powered, universal-bearing swivel bases and can rotate through a 120 degree arc on the x-, y-, and z-axes. A firing-lockout mechanism disables the phasers when the nozzles are pointed at any part of the ship's hull.

Energy for the phaser banks is supplied from the warp reaction chamber, with supplementary fusion from the impulse engine system. Utilizing power from the impulse engine system for the phasers, however, causes an inverse, logarithmic impact on impulse engine efficiency and should be used as a last resort.

Each phaser bank has a range of 250,000 kilometers and a yield of 50 Giga-joules at 100 percent efficiency. This assumes a stationary target and a stationary weapons platform. Relative motion exceeding 0,0,0 on the x-, y-, or z-axis creates degradation of range and yield directly proportional to the extent of movement.

Phaser bank compartments are flooded with a special coolant and, therefore, are not normally accessible to maintenance personnel. Should repair or maintenance become necessary, the phaser coolant can be transferred into special holding tanks to allow access. Sensors in the phaser bank compartment disable the associated phasers when this coolant level drops below a pre-determined concentration.

4.01.05 TORPEDOES

Starcruisers carry seven different classes of torpedoes, each designed for a specific mission. A standard propulsion housing and sensor package is used for each. Depending upon the class of torpedo and a pre-determined, mission-specific matrix, a portion of each ship's torpedo inventory is kept ready for immediate launch in the Port and Starboard Weapons Loading Rooms. The remainder are kept in standby mode and can be rapidly activated as necessary. The following chart indicates the total numbers of torpedoes carried and the numbers in a Ready or Standby status.

Torpedo	Total	Ready	Standby
RECTORP OKL-7-1	4	2	2
SURTORP-69-2	10	2	8
MINTORP MAR-21-3	100	10	90
ECMTORP TAC-9-4	10	2	8
SENTORP RPC-2-5	10	2	8
PROTORP DJC-11-6	250	100	150
AXN-14-7	5	1	4
TOTALS	389	119	270

Torpedo Inventory

4.01.05.01 MARK I
(Record Marker—RECTORP OKL-7-1)

The Mark I Record Marker is also known as the Flight Data Recorder. Housed in armored and shielded launch tubes (one in the primary hull, or saucer, and another in the secondary hull), these units are the "black boxes" of a starship.

These units are jettisoned from the ship under specific conditions or as directed by the Commanding Officer. Circumstances which would cause the launch of a Mark I are:

(1)   Activation of the Auto-Destruct Sequence. Once the countdown reaches .01 seconds, launch is automatic.

(2)   Any violation of the Mark I Security Interlock System. Launch is automatic.

(3)   Determination by the Main Library Computer that the ship is in imminent danger of destruction as determined by a predetermined damage matrix. Launch is automatic.

(4)   Abandon Ship stations. Launch is automatic.

(5)   Launch command by the Commanding Officer or First Officer, utilizing a special, individual-specific code. A 10-second delay between launch order and launch execution allows for a countermand code to terminate launch sequence.

Data from the Main Computer are constantly fed into the massive memory of the Mark I. Once stored, it cannot be edited, altered, or erased. The Alert Status of the ship dictates the memory load. Refer to the chart below for further information.

Alert Status	Cumulative Data Extracted	Data Limits
Green	Helm Console Telemetry Navigation Console Telemetry Communication Console Telemetry Engine Performance Data Official Logs	72 hours

Alert Status	Cumulative Data Extracted	Data Limits
Yellow	Weapons Console Telemetry Sensor Inputs Shield Status Audio/Visual Records of Main Bridge Audio/Visual Records of Auxiliary Bridge	48 hours
Red	Environmental Console Telemetry Engineering Station Telemetry Intraship Audio/Visual Records Medical Records Personal Logs	36 hours

Flight Data Recorder Data Inputs

Once launched, the Mark I Record Marker Torpedo remains in a passive mode (no external transmissions) for at least 24 hours. If hostile forces are identified in the vicinity of the unit, it will remain in a passive mode until the area is clear. At that time it begins an active search for the nearest Federation starship, starbase, planetary facility, or communications buoy. When one is found, the Record Marker emits a homing signal to aid in retrieval.

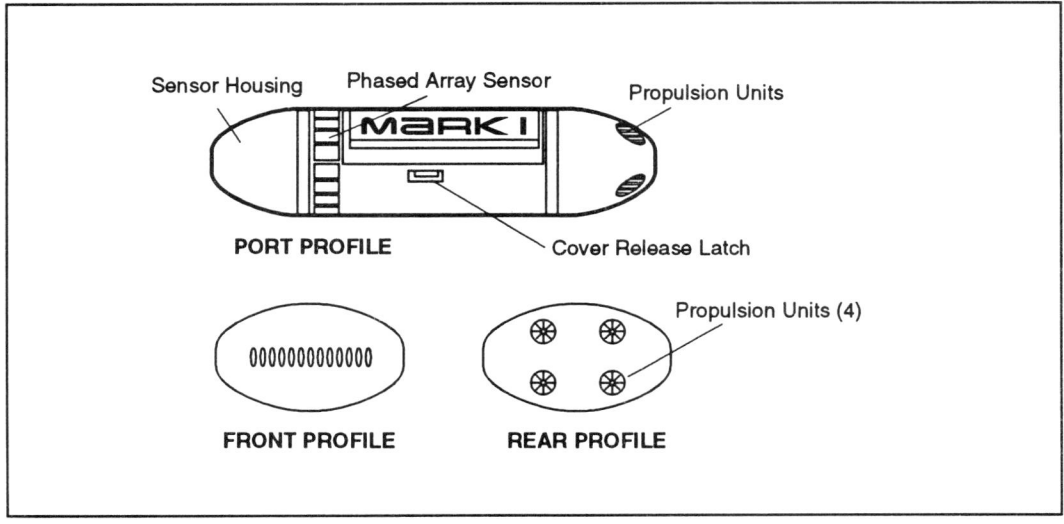

Mark I Record Marker

Dimensions

Length:	1.95 meters
Width:	0.98 meters
Height:	0.47 meters
Displacement:	98.7 kilograms

Performance

Cruising Speed:	Warp 3
Maximum Speed:	Warp 9.77 (Surge)
Range:	1.2×10^6 kilometers

Telemetry

Channels:	4,852
Output:	80 Megajoules
Sensors:	Standard Package
Additional Features:	Femto-second Data Collection
	Multi-frequency Beacon

4.01.05.02 Mark II
(Surveillance Torpedo—SURTORP-69-2)

The SURTORP-69-2 is designed to obtain military information on a specific target or defined area of space.   When launched, the Surveillance Torpedo travels at high warp to the vicinity of the target and then begins to gather data in a passive mode. Should circumstances require a more active target interrogation technique, a pre-programmed "switch" is activated which alters the basic operational mode of the torpedo from passive to active.

While it may appear that this type of mission could be accomplished by a probe, there are significant differences between the capabilities of the two. Among the differences are the following characteristics of a Mark II Surveillance Torpedo:

an extended power supply which increases Time on Target beyond the abilities inherent in a probe

forty-four phased arrays around the main housing which can operate either in a passive or an active mode are designed to obtain purely military information

an Auto-Destruct Device which causes the matter and antimatter containment bottles to vent into the torpedo body to prevent unauthorized access or to destroy or damage the surveillance target

a specialized sensor package designed to obtain specific information of strategic/tactical value on a military target (details of this package and its data parameters are classified).

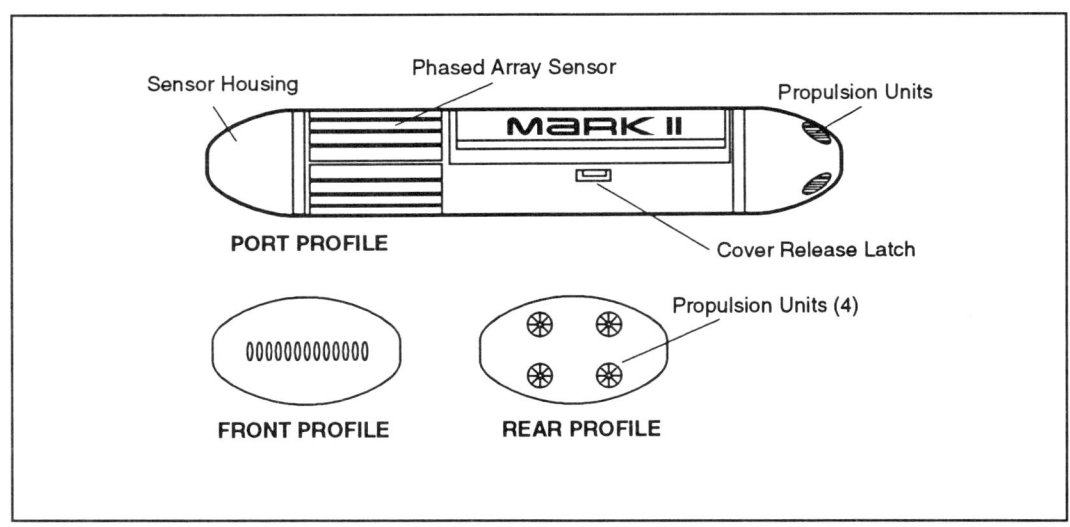

Mark II Surveillance Torpedo

Dimensions
Length:	2.75 meters
Width:	0.98 meters
Height:	0.47 meters
Displacement:	142.5 kilograms

Performance
Cruising Speed:	Warp 3
Maximum Speed:	Warp 9.77 (Surge)
Range:	1.2×10^6 kilometers

Telemetry:
Channels:	4,852
Output:	80 Megajoules
Sensors:	Military Package

Additional Features:	Femto-second Data Collection
	Multi-frequency Beacon
	Phased-array Sensors
	Auto-Destruct Device

4.01.05.03 Mark III
(Space Mine—MINTORP MAR-21-3)

Three modes of operation are available for the MAR-21-3: Active, Passive, and Shadow. Initial procedures are identical for each mode. First, the mode is selected based upon the military situation. Second, the mine is launched and, thirdly, it proceeds in a passive mode (no external transmissions) to a designated coordinate. Once in position, the selected operational mode becomes effective.

If programmed as Active, the Space Mine immediately begins a sensor sweep of its assigned area. If a target is located, the mine utilizes its ship recognition software to categorize the target. When a target is identified and classified as hostile, the mine moves at a high speed to a predetermined distance from the target and detonates.

In the Passive Mode, the mine remains at the assigned coordinates until it senses the passage of a ship within destruction range. If the ship recognition program identifies the vessel as hostile, it immediately detonates.

Shadow Mode is the least effective of the three. Once a hostile vessel is identified, the mine follows the target until one of two situations exist: the mine has insufficient fuel to continue, or multiple targets in close proximity to the mine are located. Detonation then occurs.

Space Mines cannot be moved from "standby" to "ready" status without specific orders from the Commanding Officer. Once activated, launch authorization from the Weapons Console must be verified from the Command Console.

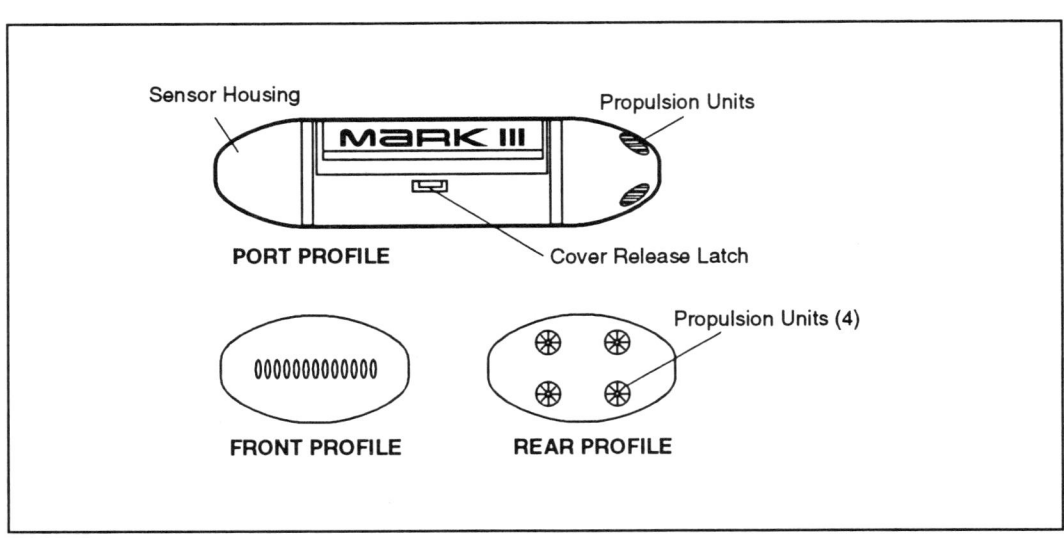

Mark III Space Mine

Dimensions
    Length: 1.95 meters
    Width: 0.98 meters
    Height: 0.47 meters
Displacement: 110.2 kilograms
Performance
    Cruising Speed: Warp 3
    Maximum Speed: Warp 9 (Surge)
Range: 1.2×10^6 kilometers
Telemetry
    Channels: 200
    Output: 12 Megajoules
Sensors: Standard Package
Additional Features: Ship Analysis Software
                       Variable Payload
                          (10—50 Megatons)

4.01.05.04 MARK IV
(ECM Torpedo—ECMTORP TAC-9-4)

Electronic Counter Measures (or ECM) are utilized in combat situations to degrade enemy sensor and fire control capabilities. In general, ECM Torpedoes emit bursts of electromagnetic interference at oscillating frequencies throughout the spectrum at super-high energy levels in a 180 degree area. Under ideal conditions, a target will be deprived of all automatic weapons control systems and be rendered completely "blind."

The ECM Torpedo can be launched in either the Homing or Mine mode. In the Homing Mode, the ECMTORP is directed toward a specific set of coordinates and activates as soon as it arrives. In the Mine Mode, the ECMTORP lies stationary until a hostile target is identified by its Ship Analysis software. It then becomes active.

Dimensions
    Length: 2.75 meters
    Width: 0.98 meters
    Height: 0.47 meters
Displacement: 139.8 kilograms
Performance
    Cruising Speed: Warp 3
    Maximum Speed: Warp 9.77 (Surge)
Range: 1.2×10^6 kilometers
Telemetry:
    Channels: 4,852
    Output: 80 Megajoules
Sensors: Military Package
Additional Features: Femto-second Data Collection
                      Ship Analysis Software
                      Electronic Countermeasures Suite

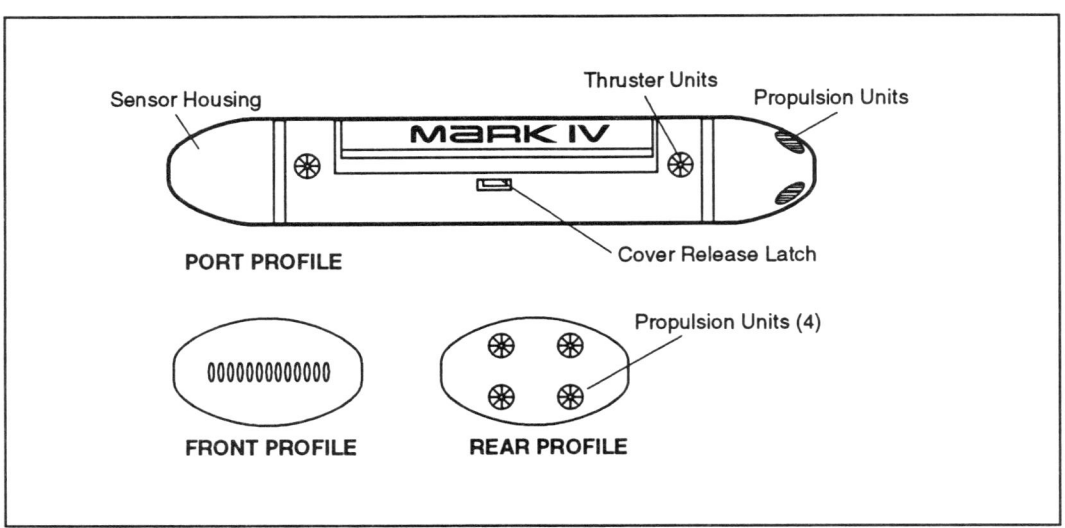

Mark IV ECM Torpedo

4.01.05.05 MARK V
(Sensor Torpedo—SENTORP RPC-2-5)

Used for long-range reconnaissance missions, the Sensor Torpedo differs from the Long-range Probe in that the Mark V is designed with the capability to attack a designated target should the reconnaissance mission be discovered. This is accomplished by an Auto-Destruct Device which causes the matter and antimatter containment bottles to vent into the torpedo body to prevent unauthorized access or to destroy or damage the surveillance target.

The Sensor Torpedo has 425 phased-array sensor discs along the lower portion of the payload section. Overlapping cones of effect give this torpedo an exceptionally sensitive data acquisition system.

Dimensions	
Length:	2.75 meters
Width:	0.98 meters
Height:	0.47 meters
Displacement:	142.5 kilograms
Performance	
Cruising Speed:	Warp 3
Maximum Speed:	Warp 9.77 (Surge)
Range:	1.2×10^6 kilometers
Telemetry	
Channels:	4,852
Output:	80 Megajoules
Sensors:	Standard Package
Additional Features:	Femto-second Data Collection
	Multi-frequency Beacon
	Multi-Phased-Array Sensors

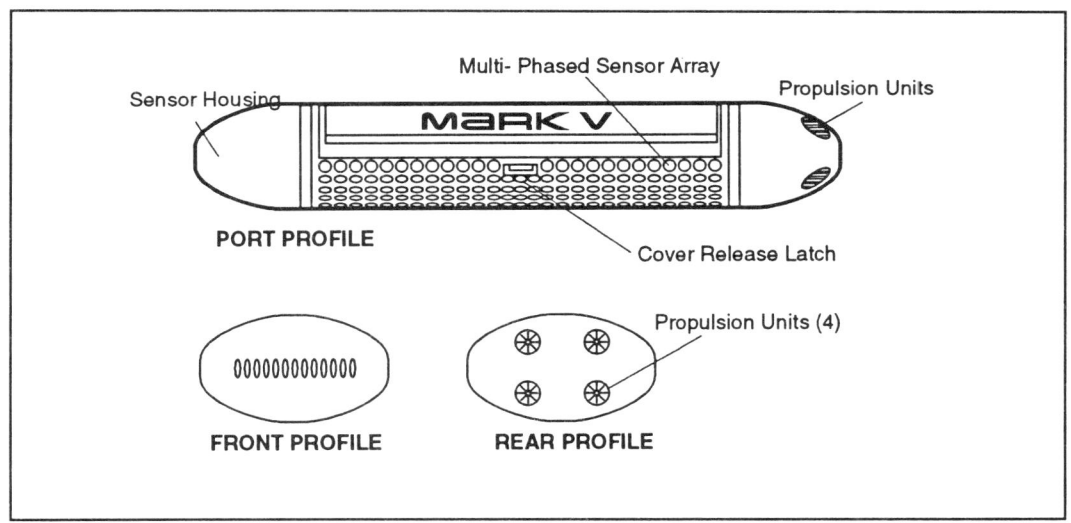

Mark V Sensor Torpedo

4.01.05.06 MARK VI
(Photon Torpedo—PHOTORP DJC-11-6)

Based on a Romulan Star Empire design, the Photon Torpedo is the most powerful weapon in the Federation arsenal. Its matter-antimatter warhead can be programmed for either a shaped or blunt pattern.

The shaped pattern is used in initial salvos to penetrate and disable target shields. Secondary salvos, if required, utilize a blunt pattern to batter the target until it is categorized as a low-threat target or is destroyed.

Instead of incorporating heavy antimatter particles (such as anti-protons and anti-neutrons), the DJC-11-6 warhead is composed of anti-protons, which create a more powerful shock wave and increases potential damage to the target.

Dimensions:
- Length: 2.75 meters
- Width: 0.98 meters
- Height: 0.47 meters

Displacement: 140.3 kilograms

Performance:
- Cruising Speed: $.96c$
- Maximum Speed: Warp 9.8

Range: 1.2×10^6 kilometers

Telemetry:
- Channels: 300
- Output: 20 Megajoules

Sensors: Standard Package

Additional Features: Variable Payload (10—50 Megatons)

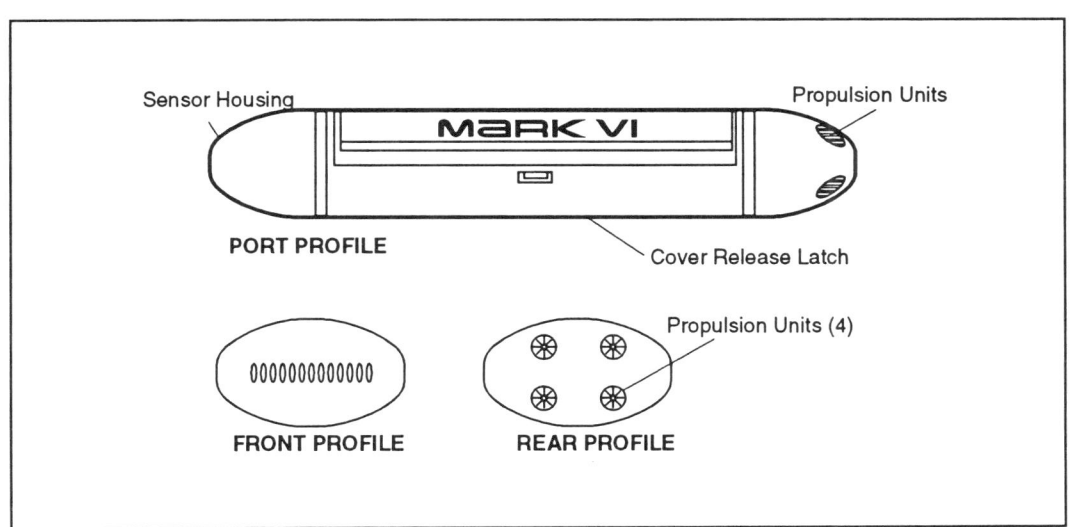

Mark VI Photon Torpedo

4.01.05.07 MARK VII
(Vessel Simulator Torpedo—AXN-14-7)

The AXN-14-7 can be programmed to emulate any one of a number of Starfleet vessels in all aspects except visual. These sensor doppelgänger-effect torpedoes can be used to confuse the enemy's tactical picture and degrade the quality of tactical or strategic decision making.

An example of one tactical situation in which the AXN-14-7 can be of value is in the case of a superior enemy force moving to engage an inferior Starfleet force. By launching an appropriate number of Vessel Simulation Torpedoes, the numbers of Starfleet ships involved can appear to be much larger. Faced with such an array of ships, the enemy commander may well decide that an engagement is unwise and therefore withdraw.

The AXN-14-7 Vessel Simulator Torpedo is often used in conjunction with the Mark IV ECM Torpedo to strengthen the effects of these types of maneuvers.

Mark VII Vessel Simulator Torpedo

Dimensions
    Length: 2.75 meters
    Width: 0.98 meters
    Height: 0.47 meters
Displacement: 138.2 kilograms
Performance
    Cruising Speed: Warp 3
    Maximum Speed: Warp 9.77 (surge)
Range: 1.2×10^6 kilometers
Telemetry
    Channels: 4,852
    Output: 180 Megajoules
Sensors: Standard Package
Additional Features: Femto-second Data Collection
                     Multi-frequency Beacon
                     Simulator Array
                     Vessel Simulation Software

4.02 AUXILIARY CRAFT

Four different classes of auxiliary craft are carried on each Starcruiser: Standard, Medical, and Aquatic shuttles; and WorkBees. Each Starcruiser's NCC number is placed before the shuttle's hull number to identify the shuttle craft's "mother ship." For example, a *Galileo*-class Standard Shuttle from USS STEPHEN W. HAWKING (NCC-8899) would be identified as 8899/4.

The hull numbers, classes, models, and types of these craft assigned to each GALILEO-class Starcruiser are shown below.

Number	Class	Model	Type
/1	*Galileo*	MK-III	Standard
/2	*Galileo*	MK-III	Standard
/3	*Galileo*	MK-III	Standard
/4	*Galileo*	MK-III	Standard
/5	*Galileo*	MK-III	Standard
/6	*Galileo*	MK-III	Standard
/M1	*Hippocrates*	MK-IV	Medical
/A1	*Cathark*	MK-I	Aquatic
/B1	*WorkBee*	MK-II	WorkBee
/B2	*WorkBee*	MK-II	WorkBee

Starcruiser Shuttlecraft Types

4.02.01 STANDARD SHUTTLE

Each Starcruiser carries six MK-III, *Galileo*-class standard Starfleet shuttles. These shuttles are used for personnel and equipment transfers when the use of transporters is impractical or contra-indicated. Each can carry up to 10 personnel, including a pilot and a co-pilot (usually the Chief Mechanic assigned to the particular shuttle). With two SW08/1-4AX warp drive nacelles, the shuttle is capable of speeds up to Warp 4.

Designed primarily for personnel, the Standard Shuttle is capable of transporting up to 0.79 metric tons of cargo. When the shuttle is used for this purpose, the interior passenger seats are removed, and the large cargo door on the stern is used for loading.

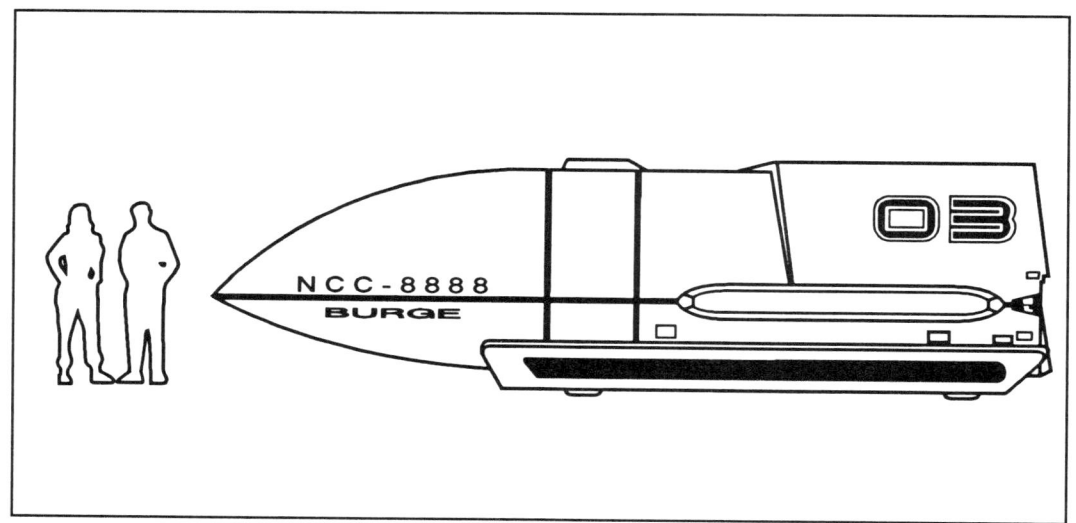

Standard Shuttle

Dimensions
    Length: 8.75 meters
    Width: 2.53 meters
    Height: 1.83 meters
Displacement
    Light: 3.10 metric tons
    Standard: 3.38 metric tons
    Heavy: 3.89 metric tons
Performance
    Impulse Engines
        Engine Output: 7.8×10^8 Megajoules
        Maximum Cruising: *.99 c*

Acceleration Rates
$\quad$ 0.00—0.25c $\qquad$ 0.137 seconds
$\quad$ 0.25—0.50c $\qquad$ 0.206 seconds
$\quad$ 0.50—0.75c $\qquad$ 0.275 seconds
$\quad$ 0.75—0.99c $\qquad$ 0.343 seconds
Warp Engines
$\quad$ Engine Output: $\qquad$ 1.2×10^7 Megajoules
$\quad$ Speeds
$\qquad$ Optimum: $\qquad$ Warp 2
$\qquad$ Safe Cruising: $\qquad$ Warp 4
$\qquad$ Maximum: $\qquad$ Warp 4.2
$\quad$ Acceleration Rates
$\qquad$ Warp 1—Warp 2: $\qquad$ 2.450 seconds
$\qquad$ Warp 2—Warp 3: $\qquad$ 2.987 seconds
$\qquad$ Warp 3—Warp 4: $\qquad$ 5.684 seconds

4.02.02 AQUATIC SHUTTLE

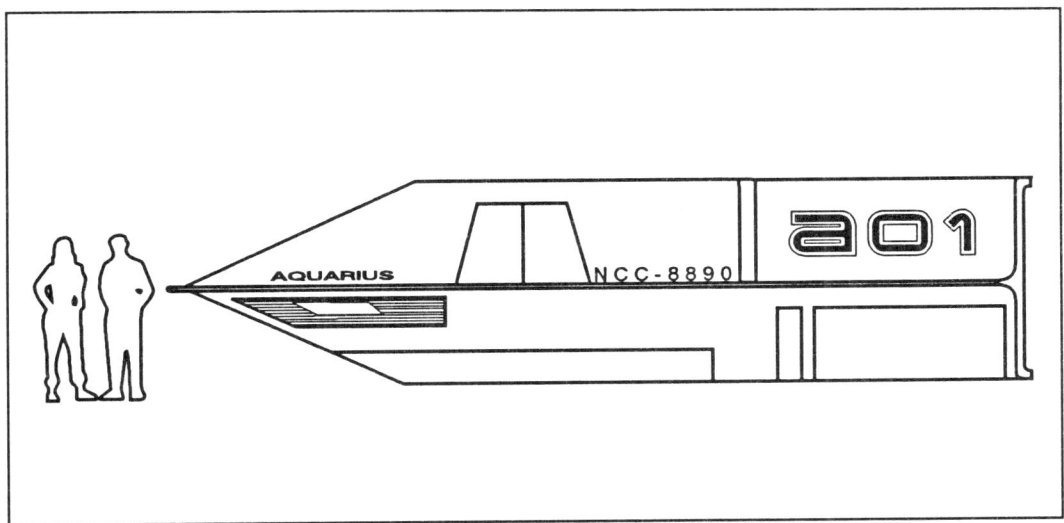

Aquatic Shuttle

The Aquatic Shuttle's hull is specially reinforced to withstand both the negative pressure of space and the positive pressure of a fluid environment.

An upper hatch connects to an air lock, allowing crewmembers to leave the shuttle while it is submerged. Specimens may be transported with tanks designed to maintain an appropriate, natural habitat.

A tractor beam capable of towing 5.1×10^2 metric tons is included as standard equipment. Should an emergency arise, an Aquatic Breathing Apparatus for each crewmember is stored internally.

Dimensions
  Length: 9.74 meters
  Width: 4.86 meters
  Height: 2.48 meters
Displacement
  Light: 3.52 metric tons
  Standard: 3.76 metric tons
  Heavy: 4.11 metric tons
Performance
  Engine Output: 7.8×10^8 Megajoules
    Maximum Cruising: $.99\ c$
    Acceleration Rates:
      0.00—$0.25\ c$ 0.137 seconds
      0.25—$0.50\ c$ 0.206 seconds
      0.50—$0.75\ c$ 0.275 seconds
      0.75—$1.00\ c$ 0.343 seconds

4.02.03 MEDICAL SHUTTLE

Because the possibility that a Starcruiser may face a situation in which out-of-the-ordinary medical demands might be required is extremely high, each Starcruiser carries *one Hippocrates*-class medical shuttle. This special shuttlecraft is utilized only for medical emergencies encountered on a planet or on a shuttle-capable vessel. Based on the *Galileo*-class standard shuttlecraft, the interior has been modified to include examination tables, diagnostic equipment, and a complete line of medical supplies.

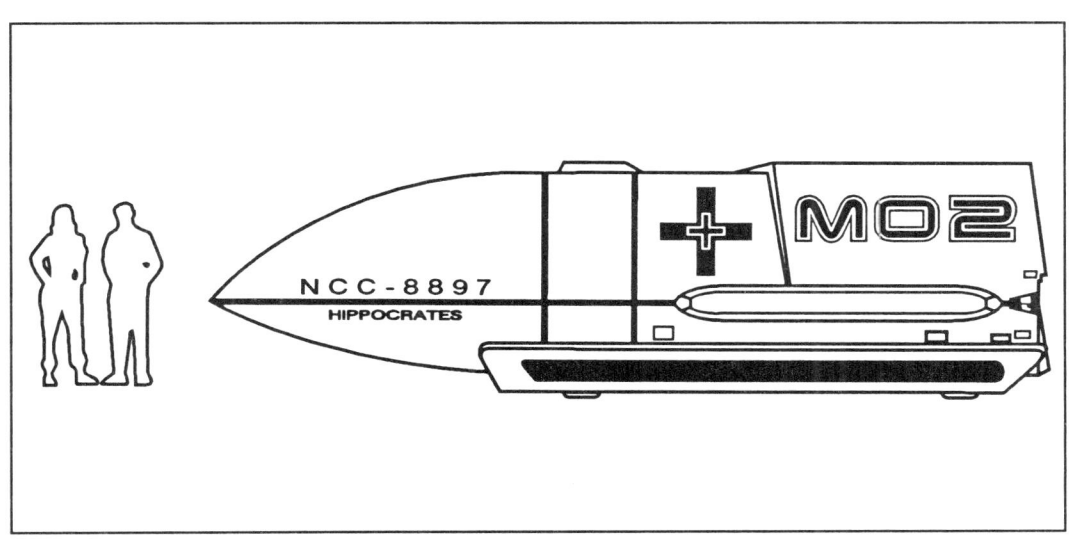

Medical Shuttle

Dimensions
    Length: 8.75 meters
    Width: 2.53 meters
    Height: 1.83 meters
Displacement
    Light: 3.10 metric tons
    Standard: 3.42 metric tons
    Heavy: 3.95 metric tons
Performance
    Impulse Engines
        Engine Output: 7.8×10^8 Megajoules
        Maximum Cruising: .99 c
        Acceleration Rates:
            0.00—0.25 c 0.139 seconds
            0.25—0.50 c 0.208 seconds
            0.50—0.75 c 0.277 seconds
            0.75—0.99 c 0.345 seconds

Warp Engines
    Engine Output: 1.2×10^7 Megajoules
    Speeds
        Optimum: Warp 2
        Safe Cruising: Warp 4
        Maximum: Warp 4.2
    Acceleration Rates
        Warp 1—Warp 2:   2.450 seconds
        Warp 2—Warp 3:   2.987 seconds
        Warp 3—Warp 4:   5.684 seconds

4.02.04 WORKBEE CRAFT

This craft is basically a stripped-down, one-man spacecraft with an extremely limited operating range. It is used when external hull repair work is required. A total of 13 different modules, either individually or in combination, can be attached to the WorkBee, depending upon the assigned tasks. These modules are stored in Repair and Maintenance Center.

The following modules, or packs, are available for use on a WorkBee:

Booster Pack: provides additional towing capacity and/or minor warp capability

Clamper Pack: adds external mechanical arms which allow for manipulation of large objects

Cutter Pack: includes a fusion cutting torch for access to or removal of portions of the external hull

Drone Pack: replaces the pilot with a independent computer to perform operations not requiring hands-on supervision

Floodlight Pack: supplies large-scale illumination for repair operations

Grabber Pack: outfits the WorkBee with a less-sophisticated mechanical arm than provided by the Clamper Pack

Heavy Booster Pack: increases towing capacity and supplies medium warp capability

Sensor Pack: enables the craft to support detailed sensor operations

Spinner Pack: allows for spot welding of hull plates and cable spooling

Survey Pack: equips the craft with the ability to perform simple survey tasks

Tow-Hitch Pack: enables a WorkBee to physically tow objects

Tractor Pack: supplies tractor beam capabilities

Welder Pack: supports precision welding in tight quarters

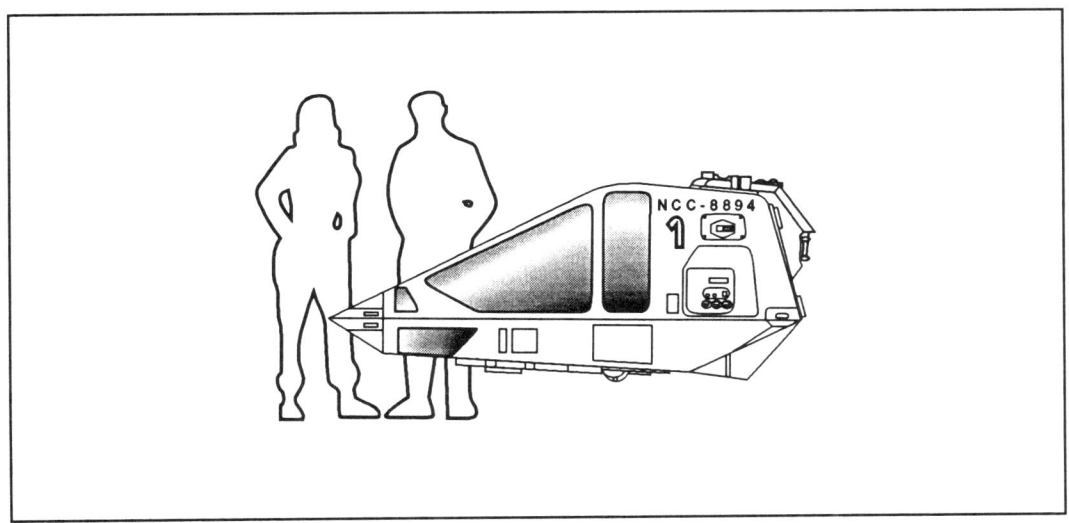

WorkBee Craft

Dimensions
   Length: 2.70 meters
   Width: 1.20 meters
   Height: 1.30 meters

Displacement
- Light: 2.12 metric tons
- Standard: 2.23 metric tons
- Heavy: 35.68 metric tons

Performance
- Impulse Engines
  - Engine Output: 9.90×10^8 Megajoules
  - Maximum Cruising: .99 c
  - Acceleration Rates:

0.00—0.25 c	15.025 seconds
0.25—0.50 c	22.538 seconds
0.50—0.75 c	30.050 seconds
0.75—0.99 c	37.563 seconds

4.03 BACKUP SYSTEMS

Redundancy was a major consideration in the design process of the GALILEO-class Starcruiser . A three-level matrix was used to determine redundancy requirements. Each shipboard component and/or system was categorized as either Level 1 (Critical), Level 2 (Important), or Level 3 (Non-essential). Each level was further classified as to the priority of the system concerned.

Maximum redundancy was applied to all Level 1 components, while Level 2 and Level 3 components qualified for decreasing attention. To illustrate this procedure, the following chart indicates the redundancy designations of selected shipboard components:

Level	Item
1A	Life Support Systems
3F	Recreation Facilities
1C	Fire-Control Systems
2B	Food Replicators
1D	Medical Support Systems

Redundancy is accomplished in several ways, depending upon the requirements of the specific component or system. In some cases multiple, duplicate systems were installed; in other cases, alternate circuits were designed to assume the particular task should a failure occur. If a system loses its main power supply, Engineering is equipped to reroute those circuits through any number of power transfer conduits without significant impact on equipment already utilizing that conduit.

Maximum redundancy is applied to life-support systems. A brief explanation of the back-up life-support provided to the Main Bridge is provided as an example of the importance attached to this system.

Atmosphere and artificial gravity are supplied to the Main Bridge by two independent circuits. Each of these is isolated from all other shipboard life-support circuits and is physically separated from the other. Should the primary circuit fail the secondary is automatically activated and priority notification of the circuit failure is made to Main Engineering personnel so that repair efforts can be instituted immediately.

Should the secondary circuit fail, a tertiary system to provide gravity and atmosphere to the Main Bridge is installed. The components of this self-contained life-support system are located on the Main Bridge itself, in the space between the two turbolifts. A dedicated power supply composed of long-life batteries provides sufficient energy for this system to operate for up to seven days.

4.04 COMPUTER SYSTEMS

Six main components make up the Starcruiser's computer system: a Main Memory Unit; a main computer processor (also known as the Main Library Computer); and four computer sub-units, one each for Engineering, Fire Control, Operations, and Science. A review of the simplified Computer Interface Schematic on the next page provides an overview of the complete system.

The Main Memory Unit, which is actually a part of the Main Library Computer, contains all data utilized or generated by the secondary computers.

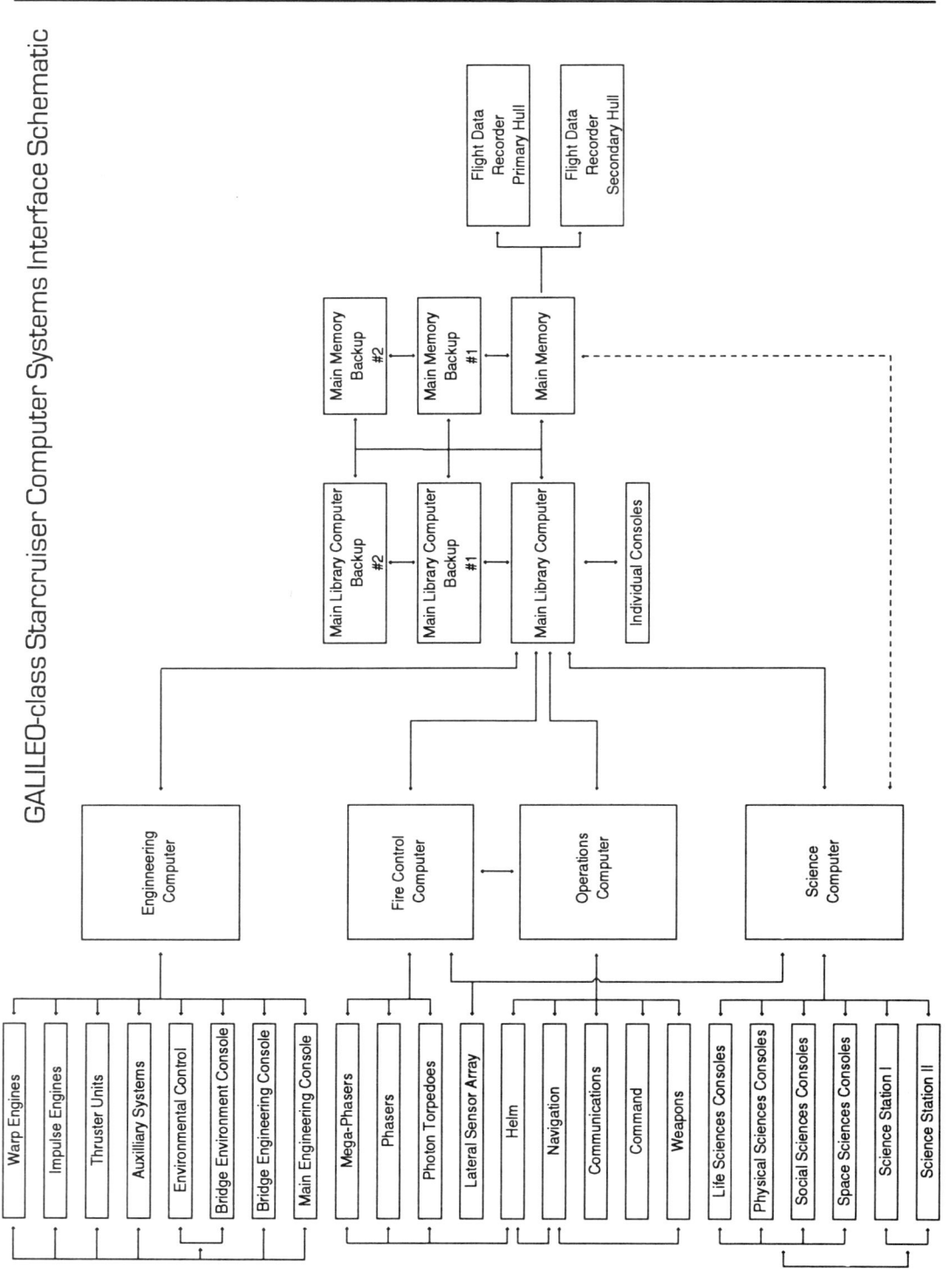

GALILEO-class Starcruiser Computer Systems Interface Schematic

Two complete backup units replicate its contents, with file updates constantly made in real time. Total capacity of each of the three units is 500 tera-bytes, with a refresh rate of 1.565 pico-seconds.

All data transmission between the several sub-systems is bi-directional with the exception of Record Marker inputs. For a more complete discussion of this particular process, see Page 28.

4.04.01 MAIN LIBRARY COMPUTER

The Main Library Computer (MLC) is also known as the Central Processing Unit. Although it provides access to the Main Memory for all individual computer consoles throughout the ship, its primary function is to control the integration of the four secondary computer systems. In this regard, it acts as a gatekeeper, routing inquiries, access, input, and data among the sub-systems and Main Memory input and output.

As an example of this process, assume the EOOW is at the Main Engineering console and asks for data on warp engine performance for the past eight hours. When this signal reaches the MLC, its priority is matched against other signals present in the MLC buffer, or queue. Based on the relative priority level, the MLC will either assign the inquiry a place in line, or it will route the request directly to the unit's's main processor. If routed to the Main Processor, the MLC will recognize the signal, access the appropriate area of Main Memory, and provide the requested data.

The MLC's multiple parallel processing capabilities enable it to respond to any request, even a low-priority one, within 15 nano-seconds. This high computation rate results in the appearance of instantaneous access to all consoles and individual terminals on the ship.

Designed and built by Monolithic Memories of Vulcan (40 Eridani), the "Mega Monolithics" is based on recent technological breakthroughs in Monolithic Memories' Research and Development section. It's unique design makes the computer interface system aboard the GALILEO-class Starcruiser the most responsive currently available in Starfleet.

The Main Library Computer contains virtually every item of information which might conceivably be required or desired under any envisioned situation during an extended mission. Second in capabilities only to Memory Alpha, the Main Library Computer contains such items as:

the text of over 20 million books;

50 million research papers in all areas of scientific study;

physical, historical, economic, and social data on every planet in the Federation;

every survey report submitted to Starfleet;

a complete copy of the 1,000-volume, 207th edition of the Encyclopedia Galactica; and,

holographic images of fine art and sculpture from both modern societies as well as archaeological discoveries.

Since classified materials are included in the memory of the Main Library Computer, each individual on the ship has been assigned a Billet Sequence Code which is cross-referenced by the computer to indicate that person's security access level. When signing onto a terminal, the user must enter a Billet Sequence Code and password to safeguard sensitive information.

4.04.02 ENGINEERING COMPUTER

As part of the Design Specification Package, Monolithic Memories of Vulcan was faced with the task of delivering a computer which would automate all functions in five different areas: warp engines, impulse engines, thruster units, auxiliary systems, and environmental controls. Previous designs treated each of these responsibilities as separate and unrelated taskings. However, the complexity and interconnectivity of engineering systems aboard the GALILEO-class Starcruiser precluded such a standard approach.

The scientists involved in the research and development for the engineering computer, many on loan from the Daystrom Institute, approached the process from a revolutionary viewpoint. Instead of five different control systems, an unified concept was adopted. This resulted in the delivery of an integrated control system which treated all inputs as different aspects of the same executive routine.

As a result, the "Montgomery II" is the first computer system which analyzes the effects of a change in one sub-system on another sub-system and automatically modifies operational parameters as necessary. Hierarchial review of real-time paradigms results in equilibrium throughout the five areas, with out-of-specification operating conditions reported immediately to the EOOW. Alternatives are presented along with cascade effects for optimum decision making.

4.04.03 Fire Control Computer

The fire control computer, the least sophisticated computer system aboard the GALILEO-class Starcruiser, was designed as a restricted-capability unit. Located in the Torpedo Bulge at the apex of the Weapons Bridge, the "FasTrac" handles only five functions:

Threat identification
Threat analysis
Target tracking
Weapons assignment
Weapons firing

A dedicated sensor package provides input on a target specified by the Weapons Console operator. These inputs are correlated with data in the MLC and identification made of the target. The Weapons Console operator is advised of the status of the target as either "friend" or "foe." Specific identification of the target (type of target, name, relative position, course and speed, and projected closest point of approach) is also provided.

Simultaneously, an analysis of the relative threat posed by the target is projected. This threat analysis uses data from the MLC on speed, man euverability, weapons capabilities, shield capacity, and known tactical scenarios, to categorize the target's threat level on a scale of 1 to 10. Should multiple targets be identified, each threat level is identified in priority order.

Real-time updates to course, speed, threat level, and projected courses are supplied for all designated targets. Any changes to weapons or shields status is automatically highlighted as the fire control computer tracks each individual target.

Based on the threat analysis and other critical determinations (including own-ship's weapons, power consideration, damage level, and tactical input from the Weapons Console operator), an appropriate weapon selection recommendation is made. These recommendations can be accepted or modified as necessary as dictated by tactical considerations.

Once a target is selected and the weapons mode is determined, the fire control computer will take over all firing functions in order to effect the maximum possible damage. Should manual control be instituted, recommendations will continue to be presented. A return to automatic control can be accomplished at any time. Battle Damage Assessment data on each selected target is updated on an as-occurring basis.

If the fire control computer is damaged, its functions can be assumed by either the Operations or the Science computer.

4.04.04 OPERATIONS COMPUTER

Five major sub-systems are controlled by the Operations Computer: Helm, Navigation, Communications, Command, and Weapons. Input/output nodes are bi-directional as indicated in the Computer Interface Schematic. The Lateral Sensor Array also provides continual input.

Should the Operations Computer fail or require maintenance, control of the sub-systems is transferred automatically to the Science Computer. Functions of the Fire Control Computer can be assumed by the Operations Computer if necessary.

4.04.05 SCIENCE COMPUTER

Second in capabilities only to the MLC, the Science Computer assumes tasks allocated to the Operations Computer during periods of maintenance or system failure. Main control of this system's assets resides in Science Station 1 on the Main Bridge; Science Station 2 is an alternate.

The consoles in Life, Physical, Social, and Space Sciences receive computational support on a prioritized basis; however, Science Station 1 and Science Station 2 (respectively) have the highest priority level.

If the Main Library Computer and its two backup systems fail, all functions of the MLC are taken over by the Science Computer. A measurable degradation of response time and simultaneous computational capabilities will occur, and, should this situation arise, non-essential consoles are automatically locked out of the MLC loop.

4.05 HULL CONSTRUCTION

The outer hull is a combination of six relatively new metals: Durolithium, Polyhedrium, Humbrolite, Mychromium, Paumerium, and Overticite. Although it does not provide complete protection, this unique combination has proved to be highly resistant to detection by every known sensor system.

The two center layers, composed of Humbrolite and Mychromium, are the thickest, and are in the form of three-quanta-module Mites. Humbrolite is used for two A Quanta Modules, while Mychromium is used for the single B Quanta Module. Utilizing this simplest minimum-limit case of allspace-filling polyhedra concept for the base construction increases hull resistance to inertial forces.

The two outer layers, composed of Durolithium and Polyhedrium, are reverse polarized. One result of this polarization is layer-bonding at the atomic level, which increases tensile strength logarithmically. Another result is to reduce (like the center two layers of the hull) the effects of inertia on the overall hull structure.

Paumerium and Overticite make up the two interior layers of the hull. Paumerium is a derivative of Transparent Aluminum, but possesses double the rigidity of that compound. Overticite is permeable, which absorbs electromagnetic radiation throughout the spectrum. It is this layer that contributes the majority of the hull's sensor resistibility.

As is now customary in Starfleet vessels over 90,000 metric tons, no thermocoat has been applied to the hull.

4.05.01 PRIMARY HULL

To provide sufficient space to meet the Design Specification Package requirement for extensive science laboratory facilities, expanded primary hull dimensions were a necessity. However, the standard Starfleet design could not be modified to provide these additional compartments without exceeding the displacement limitations.

A number of new saucer designs were created and discarded during the design process as either negatively affecting the Warp Field Profile, exceeding the displacement limits or, in at least one case, not being "aesthetically pleasing." The final approved compromise resulted from a combination of an existing design and a modification of the existing, standard saucer dimensions.

The extended-hull concept initially used on the KNOX-class Frigate and perfected with the AVENGER-class Heavy Frigate was utilized as the basic design for the primary hull. Although interior volume was increased, the additional displacement fell within specifications. The extended hull also allowed all scientific laboratories to be concentrated in one area, which reduced computer interface difficulties as well as power conduit problems.

Both the dorsal and ventral curvatures of the primary hull were changed to convex. This modification also increased internal volume without significantly affecting displacement or Warp Field Profiles.

One unintended consequence of this design was to make available more volume dedicated to living quarters. It allowed the size of staterooms to be increased and thereby contributed to the standard of living for crewmembers subjected to long voyages.

4.05.02 SECONDARY HULL

Several factors impacted the design of the Starcruiser secondary hull. Among these were the increased stress and dead weight of three warp engines, a larger dorsal resulting from the extended primary hull, larger cargo holds and replicator material banks, and a larger Flight Deck facility to support the number of shuttles to be carried. Modifications to the typical secondary hull design were made to allow for these critical specifications.

Instead of the circular cross section, an ellipsoid was used for the basic shape. At the same time, the external dimensions were increased. This combination greatly enlarged the area of the secondary hull. Computer simulations revealed that the new shape actually contributed to reducing power consumption throughout the Acceleration/Energy Index without impacting the Warp Field Profile.

4.05.03 DORSAL

The traditional two-hulled starship utilized the dorsal as little more than an interconnector. However, the size of the primary hull and the secondary hull, along with the third engine, required an attendant increase in the dimensions of the dorsal. This enlarged dorsal effectively doubled its capacity, allowing the installation of vital components and systems used by both hulls, with a decrease in the number of complexity of circuits and conduits required throughout the ship.

The majority of the VIP quarters were placed in the dorsal along with a VIP-only lounge and a conference room so that the presence of VIPs would not impact normal shipboard routine.

4.06 PROPULSION SYSTEMS

Starships have three distinctly different propulsion systems. Each is designed to provide the necessary speed to satisfy a certain set of requirements and/or control parameters.

Faster than light: used to traverse large distances between point of origin and destination, usually one star system (or spatial coordinate) to another; speeds range from c up to Eugene's Limit.

Sublight: utilized when high level of speed is not essential or where faster-than-light speeds are inappropriate (such as within a star system); speeds range from $0.001c$ to $0.999c$.

Sub-impulse: utilized when minute changes to pitch, yaw, or roll are required (such as station keeping, orbit maintenance, or docking maneuvers); speeds range from 0.001 meters per second to 1,000 meters per second.

Warp engines provide faster-than-light capabilities while the impulse engines operate at sublight speeds; sub-impulse is provided by thruster units.

4.06.01 THRUSTER UNITS

The high displacement and attendant mass/inertia co-efficients of the GALILEO-class Starcruiser presented a challenge to Scarbak Propulsion Systems of Cairo, Earth (Sol), that was awarded the thruster unit construction contract.

An off-the-shelf thruster unit, normally installed in a starship, was incapable of providing the level of finesse required to maneuver the Starcruiser in close quarters. By integrating all the thruster units together into a single system, the first step was taken to control the 305,000 metric ton ship. The second stage was to modify the thruster units to provide a regulated series of power bursts rather than a continual flow of energy. In addition, the numbers of thruster units located on the ship's secondary hull were doubled, enabling more positive control over pitch and yaw.

4.06.02 IMPULSE ENGINES

For sublight travel, the "Power Six-Pack" multi-ported impulse system built by Klorati Drives of Tellar, 61 Cygni, provides the ship with a fuel-efficient, almost maintenance-free, means of propulsion.

The impulse engine system is composed of six thrust emitters and six associated fusion reactors. Each of the thrust emitter ports is numbered beginning with the port inboard emitter as Number 1 and the starboard inboard emitter port as Number 2. The center portside emitter port is Number 3 and the center starboard emitter port is Number 4. Number 5 is the port outboard emitter port and Number 6 is the starboard outboard emitter port.

Depending upon the desired speed, sequential emitter ports are activated in numerical order. The chart below indicates the speed available by activation of each impulse engine and its associated emitter port and fusion reactor.

Impulse Emitter	Minimum Speed	Maximum Speed
1	$.001c$	$.167c$
2	$.168c$	$.334c$
3	$.335c$	$.500c$
4	$.501c$	$.667c$
5	$.668c$	$.834c$
6	$.835c$	$.999c$

Impulse Engine Performance

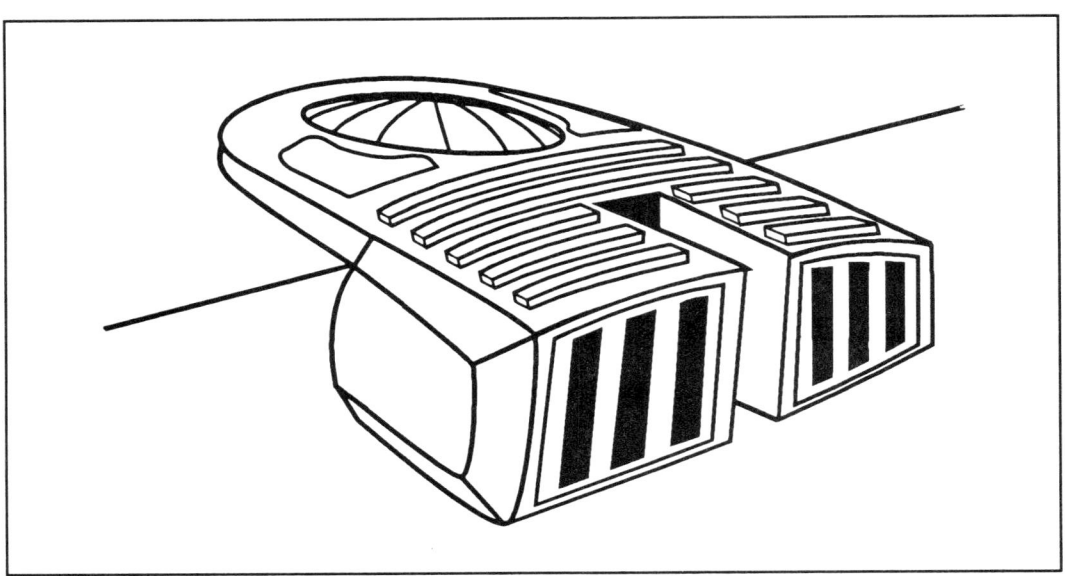

Power Six-Pack Multi-ported Impulse Engines

4.06.03 WARP DRIVE

As defined in the Design Specification Package, the Starcruiser was to possess the capability to be totally self-supporting for periods of up to five years. This would require a more fuel-efficient warp drive and impulse engine systems than were available. Based on projected displacement figures (284,532.0 metric tons light; 305,000.0 metric tons standard; and 340,528.0 metric tons full) the Warp Engine Efficiency Factor (WEEF) would have to be between 0.82 and 0.85.

Leeding Engines, Ltd., of Sydney, Earth, developed the SY-71/15AC Inter phased Warp Engine in response to this requirement with a WEEF of 0.835 and a sustained power rating of 2.16×10^{15} Megajoules. By interphasing the individual warp engine outputs, both dilithium crystal degradation and matter-antimatter consumption were reduced by approximately 18 percent.

Interphase refers to the technique of aligning the output field waves of one warp engine with another, reinforcing both warp signatures and resulting in less power required for a given mass versus speed matrix. The three-engine design (reminiscent of the STAR LEAGUE-class Dreadnought) increased this reinforcement and made available an energy surplus for use by other shipboard systems.

One warp engine is mounted on a support pylon atop the secondary hull on the centerline. Each of the other two warp engines is on a horizontal support pylon originating from the base of the "dorsal" engine support pylon. Should circumstances dictate, any one or all three of the engines can be jettisoned.

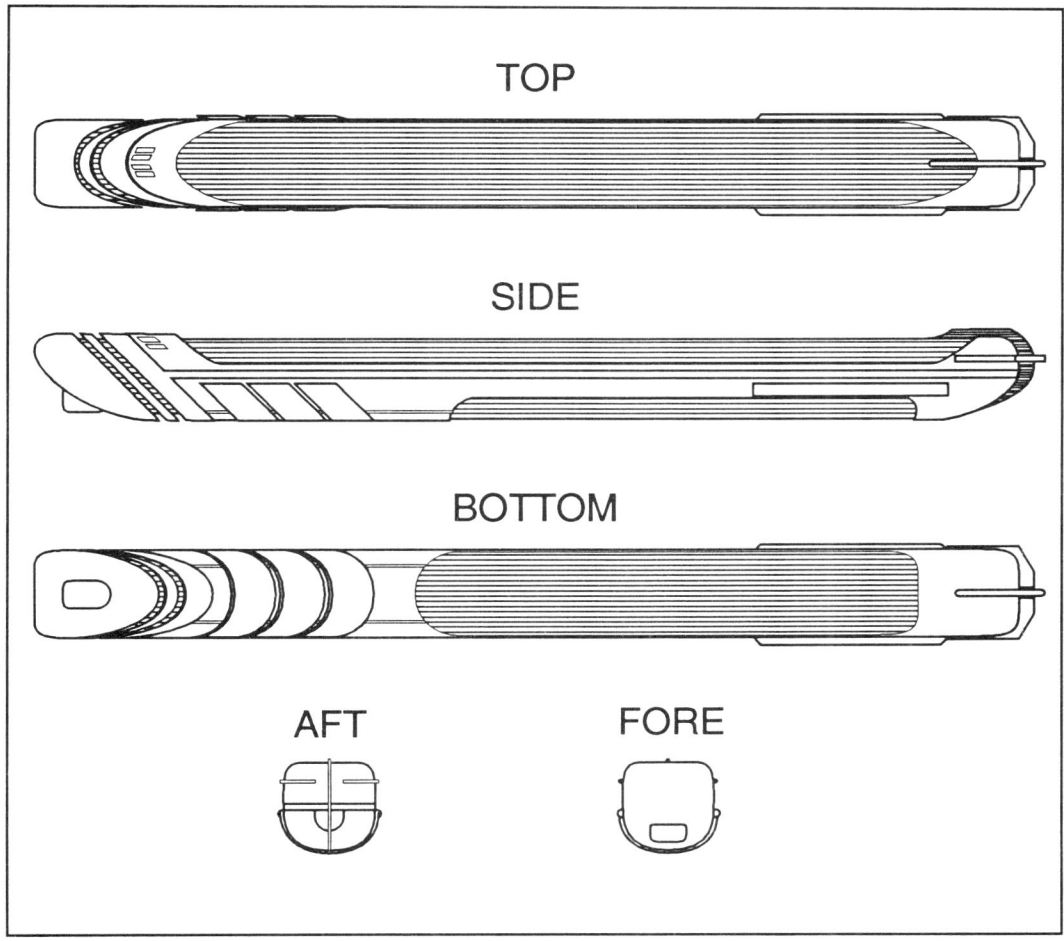

TOP

SIDE

BOTTOM

AFT   FORE

SY-71/15AC Interphased Warp Engine Nacelle

4.07 Main Bridge

The Main Bridge is the nerve center of the ship. From here, officers on watch make all tactical and strategic decisions. As is customary in Starfleet ships, the Bridge is a replaceable module.

The current configuration, the SCIEX MK-I MOD-0, was designed to reflect the scientific and exploratory mission of the GALILEO-class Starcruiser. In addition to standard station consoles (Commanding Officer, Navigation, Communications, Helm, Engineering and Environmental Controls, and Weapons), two science stations, one on each side of the Command Console, receive direct input from the sensor package. These two stations act as the control center for the Science Department during intensive research or exploratory operations.

4.07.01 Command Station

Composed of three chairs on an elevated platform, the Command Station is located in the center of the Bridge, facing the Main Viewing Screen. The center position is designated for the senior officer in charge of shipboard operations. When present, the Commanding Officer will take this position.

In the absence of the Commanding Officer, the First Officer normally assumes these duties. When the First Officer is not on the Bridge, a designated Officer-of-the-Deck (usually the Operations Officer) will act in the place of the First Officer.

When both the Commanding Officer and the First Officer are on the Bridge, the First Officer will occupy the position to the right of the Commanding Officer's center seat.

To the left of the Commanding Officer is a third position which is used by an officer or visitor at the Commanding Officer's discretion.

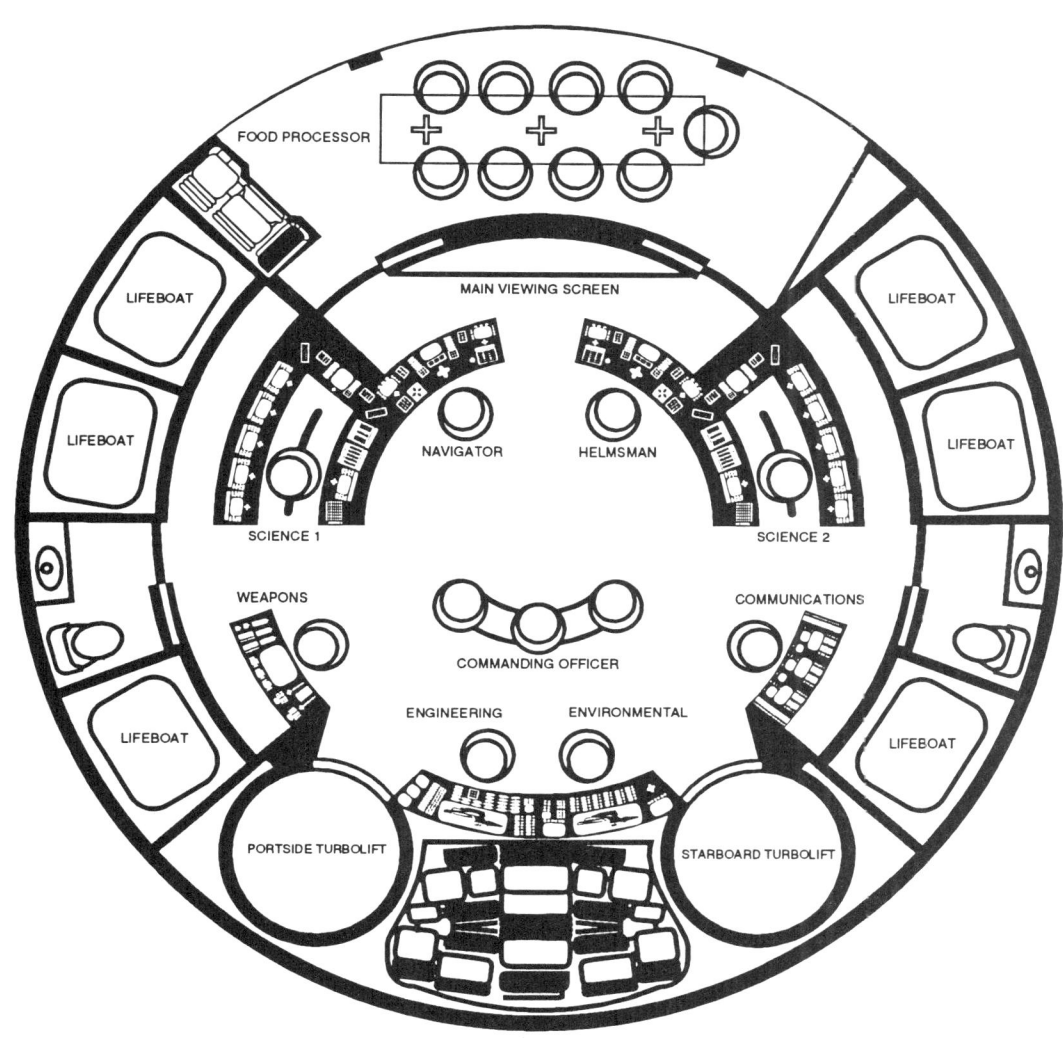

SCIEX MI-I MOD-O Main Bridge Module

A control panel between the center seat and the starboard seat provides communications access (both internal and external) and alert status (both shipwide and Security) control. A small holographic display unit can be programmed to replicate any console readout on the ship.

4.07.02 NAVIGATION CONSOLE

Just in front of, and to the left of, the Command Station is Navigation. It is continuously manned except when the ship is moored, in orbit for an extended period, or in "cold iron" status. Responsibilities of the officer assigned this position can be found on Page 119.

In the center of the Console is the Main Course Plot. A small holographic screen projects the calculated course as entered and indicates the position and course of other ships in the immediate area. The location of all star systems, starbases, and buoys is also displayed. Navigational hazards are indicated in red. A small screen on the left side of the Console provides a close-up view of any navigational hazard and can be directed to enlarge any portion of the Main Course Plot as desired.

Another small screen on the extreme right of the Console can be adjusted to show an exterior view of the ship from any externally mounted Visual Sensor Unit (VSU). If necessary, up to four different VSUs can be displayed simultaneously.

Should the Helm or Weapons consoles become inoperative, the Navigation Console has back-up weapons and thruster unit controls . In such a case, the exterior view screen is automatically shifted to provide Tactical Plot or Weapons status capability as appropriate.

Personnel filling the following positions in the Ship's Manning Document rotate through the watch schedule:

Position	Billet Sequence Code	Watch
Chief Navigator	010201	NONE
Assistant Navigator	010202	ALPHA
Navigator	010203	BETA
Navigator	010204	GAMMA

Navigation Console Personnel Assignments

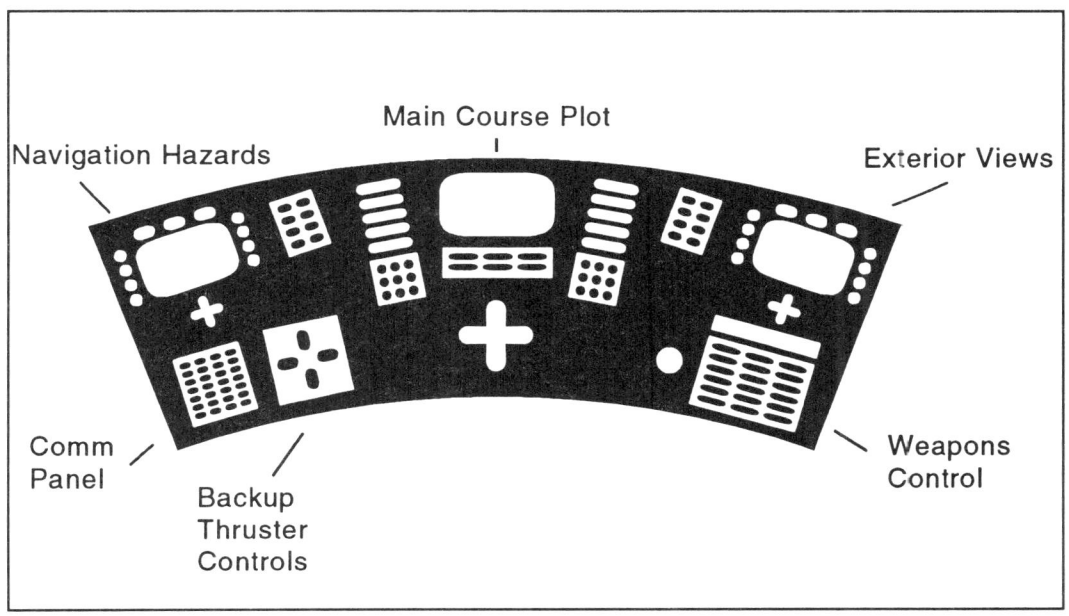

Navigation Console

4.07.03 HELM CONSOLE

Immediately to the right of the Navigation Station is the Helm Console which is manned at all times except when the ship is in drydock or in a "cold iron" status. Responsibilites of personnel manning the Helm are located on Page 120.

The layout of this console is nearly identical to that of the Navigation Console, except that there are two VSU screens, one on the top left side and another on the top right side of the console face. In the center of the console is Tactical Plot, a holographic screen which depicts all objects within 20 lightyears of the ship. Courses, speeds, or orbits (as appropriate) are displayed alongside each identification marker. Longer range views or close-up detail can be selected as desired.

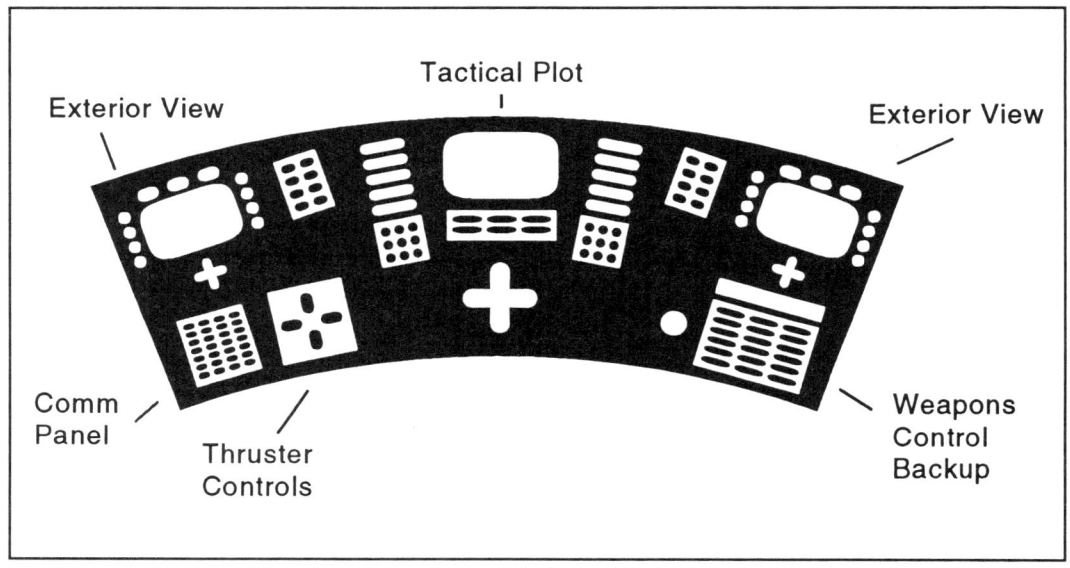

Helm Console

Main thruster controls, backup weapons controls, and a communications interface are located at the bottom of this panel.

Personnel filling the following positions in the Ship's Manning Document rotate through the watch schedule:

Position	Billet Sequence Code	Watch
Chief Helmsman	010301	NONE
Assistant Helmsman	010302	ALPHA
Helmsman	010303	BETA
Helmsman	010304	GAMMA

Helm Console Personnel Assignments

4.07.04 COMMUNICATIONS STATION

Personnel manning the Communications Station monitor all frequencies in both sub-space and sub-light levels for messages either directed to the ship, or which may have an impact on ship's operations. All outgoing transmissions are channeled through this station. Any intraship communication circuits may be overridden from this station. Control of the Main Viewing Screen resides with the officer on watch at the Communication console.

Six multi-purpose screens compose the major portion of the Communications Console. The top three are primarily used to display frequency and power characteristics of incoming subspace and sub-light signals. The bottom three provide the same information for outgoing signals.

Other information which may be displayed on these screens include communication buoy coordinates and characteristics, message authentication, receipt verification, and visual transcripts of both outgoing and incoming messages. Two small speakers can be activated if desired, although most aural services are provided by a device which is inserted into the operator's outer ear.

Controls to manage and monitor all intercom circuits are on either side of the console. Along the bottom are direct-access intercom buttons for the Commanding Officer's Suite, the Operations Officer, Engineering Officer, Medical Officer, Science Officer, Security Officer, and the Shuttlebay. An eighth intercom button can be programmed as desired.

Personnel filling the following positions in the Ship's Manning Document rotate through the watch schedule:

Position	Billet Sequence Code	Watch
Chief Communications Officer	010401	ALPHA
Assistant Communications Officer	010402	BETA
Communications Systems Officer	010411	GAMMA

Communications Station Personnel Assignments

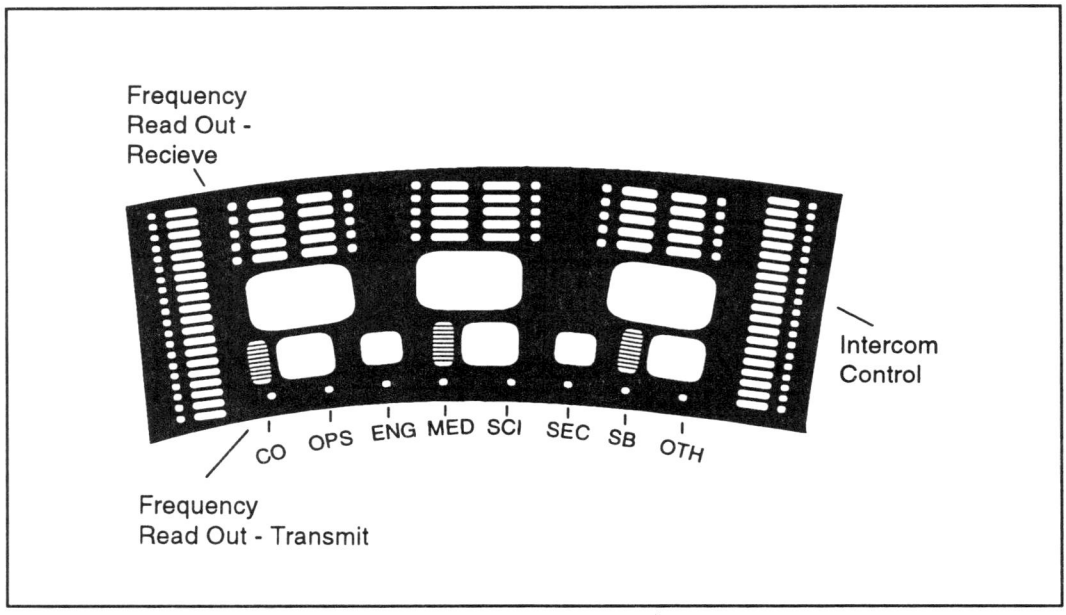

Communications Console

4.07.05 Science Stations

Because the primary mission of the Starcruiser is concentrated on scientific research and exploration, two Science Stations were included in the SCIEX

MK-I MOD-0 Bridge module. Science 1 is to the left of the Command Station and Science 2 to the right. During regular operations, only Science Station 1 is manned; Science Station 2 IS activated as required.

Each Science Station is an identical, U-shaped console with the operator's chair mounted on a glide plate to facilitate movement along the control surfaces. The base of the "U" faces the Main Viewscreen and contains a multi-purpose screen, sensor controls, and a MLC interface panel. The interior leg of the "U" provides probe configuration and launch command controls, a probe data read-out screen, and a communications panel. Routing of data to any console on the ship is accomplished by a matrix located on this section of the console.

Five screens and associated controls are on the exterior leg of the "U." Each is dedicated to a specific category of data and can receive inputs from the lateral sensor array, individual consoles throughout the ship, and tricorders. The Astrological screen provides information on objects or anomalies located in space. The Meteorological screen displays weather or climatological data. Free-fluid flow, evaporation rates, and ground water locations are shown on the Hydrological screen. Living organisms are tracked on the Biological screen, while the Geophysical screen depicts landform and surface characteristics.

Personnel filling the following positions in the Ship's Manning Document rotate through the watch schedule:

Position	BSC	Station	Watch
Chief Science Officer	030001	Science 1	ALPHA
Assistant Science Officer	030002	Science 1	BETA
Chief Life Sciences Officer	030101	Science 1	GAMMA
Chief Physical Sciences Officer	030201	Science 2	ALPHA
Chief Social Sciences Officer	030301	Science 2	BETA
Chief Space Sciences Officer	030401	Science 2	GAMMA

Science Station Personnel Assignments

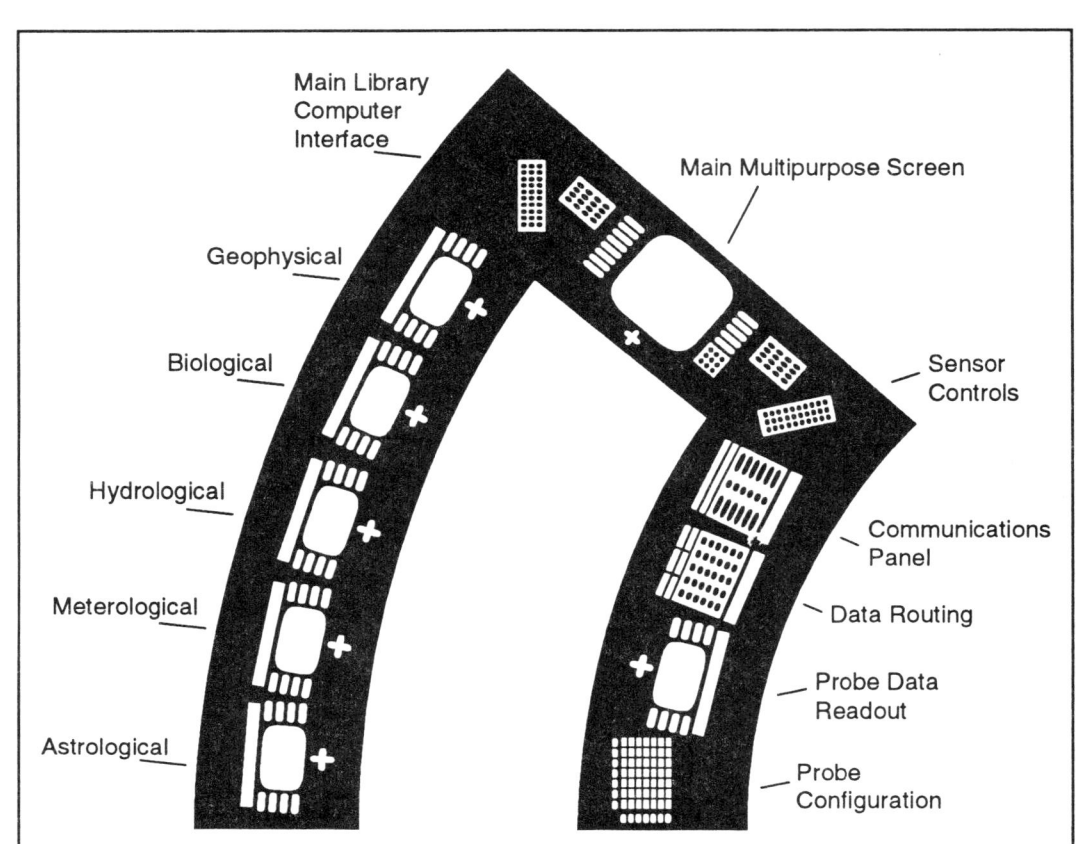

Science Stations 1 and 2

4.07.06 Engineering Station

This station, although normally unmanned under cruising conditions, is designed to monitor and override all engineering functions during Yellow and Red alerts. The Commanding Officer, First Officer, Engineering Officer, or Officer-of-the-Deck are the only individuals authorized to override engineering functions.

It is normally manned by the Assistant Chief Engineer (BSC 02003), but is often used as a training station for senior Engineering personnel to acquaint them with Main Bridge procedures.

The cut-away ship's schematic screen at the top center of this console provides real-time data on all engineering functions throughout the ship. Color key pads control not only the circuit colors of the schematic, but also enable the operator to include or exclude any conduit, circuit, or component from the display.

An Engine Performance Readout at the top right displays the current status of the engine system in use (warp, impulse, or thruster). Two programmable readouts just below can provide historical usage and conditions of any chosen engineering function, from transporters to replicators.

Engineering Interface Pads enable the operator to over ride any action taken in Main Engineering.

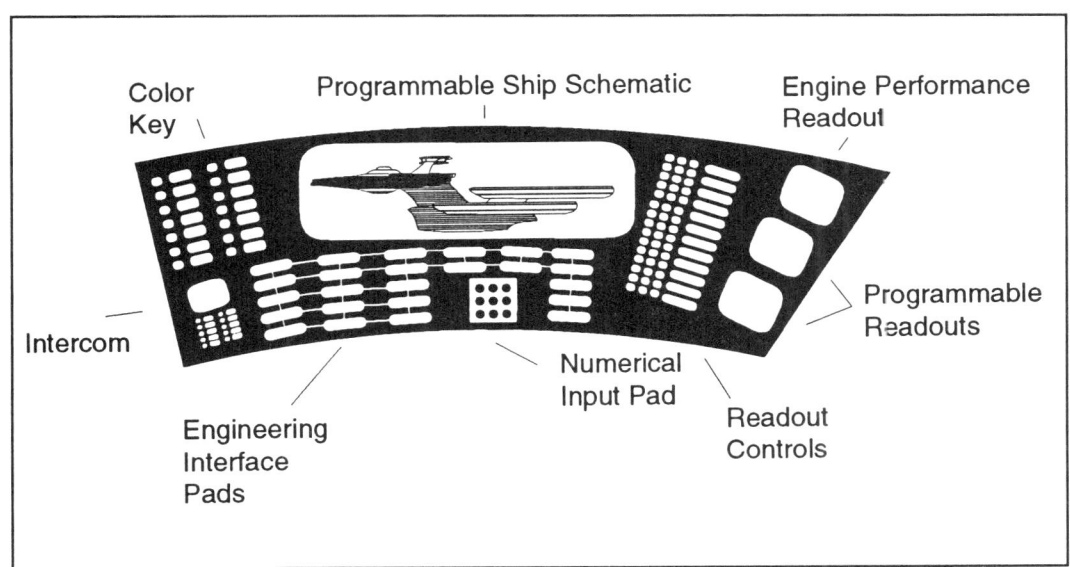

Engineering Console

4.07.07 ENVIRONMENTAL CONTROLS STATION

A large, cut-away schematic of the ship is the predominate feature of this console. The operating status of all environmental circuits and components are displayed. By using the schematic control on the lower right corner, the operator can change the schematic to display the environmental conditions of individual compartments or a specific area of the ship.

Two programmable readout screens, one at the top left and the other at the top right, can duplicate sections of the ship's schematic data, or other equipment status as desired.

Controls beneath the large schematic screen enable to operator to change the atmosphere, humidity, light, or gravity in any compartment or section of the ship.

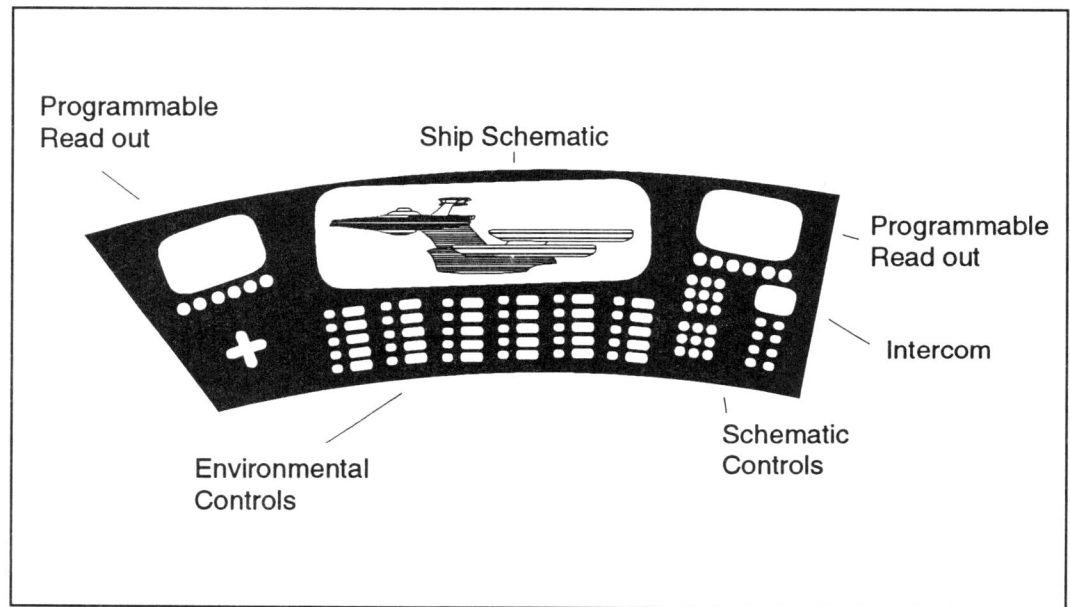

Environmental Controls Console

4.07.08 WEAPONS STATION

During Yellow and Red Alert this console is manned by an individual responsible for controlling phasers, Mega-phasers, and photon torpedoes under the direction of the senior officer present. It is normally manned by the

Chief Weapons Systems Engineer (BSC 020401). Emergency control of weapons during normal cruising operations is available at the Navigation Station.

A large, holographic Tactical Display Screen dominates the console. It enables the operator to view a desired area of space or planetary body for more accurate targeting. The level of detail provided can be specified by a bank of control to the screen's right.

Torpedoes are programmed and launched by a matrix of panels just above the Tactical Display Screen controls. Three small holographic screens at the top center of the console can be used as desired to provide additional data input to the operator.

Mega-phasers are manipulated, aimed, and fired by a series of touch-sensitive pads at the top, left side of the console. Just below this area are the controls for the standard shipboard phasers.

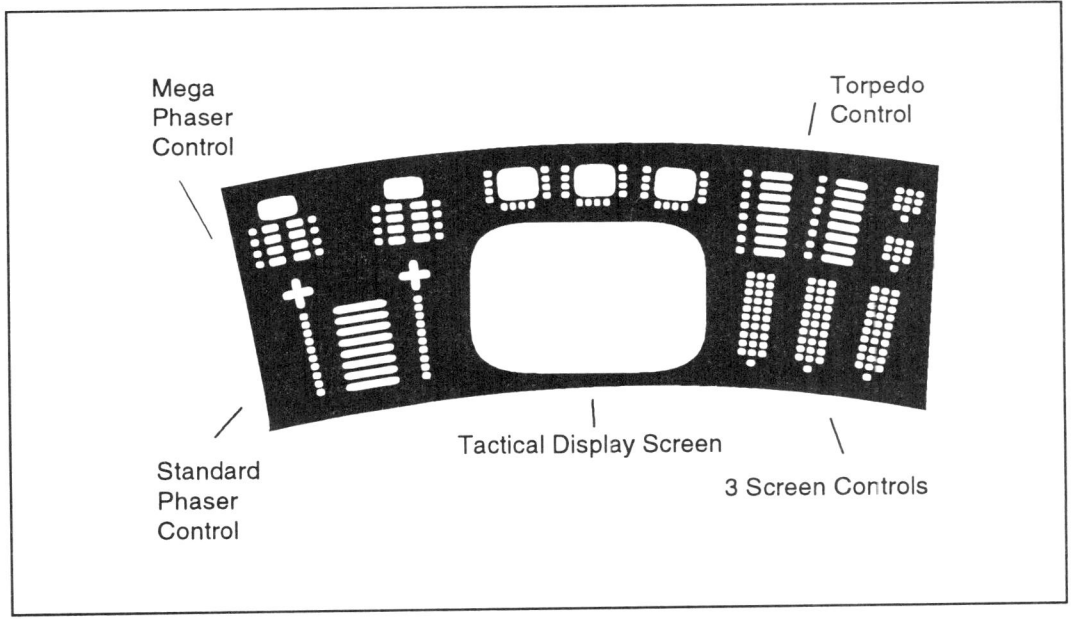

Weapons Console

4.07.09 BRIDGE BRIEFING ROOM

In a departure from previous Main Bridge designs, the Captain's main briefing room was moved to an area forward of the Bridge itself. This location is more convenient for operational, strategic, tactical, and decision-making discussions.

A large conference table takes up the majority of the briefing room. Each of the nine places around the table has a control panel which operates a holographic display unit in the center of the table and viewscreens on the aft bulkhead. Computer interface is also available from these control panels.

Three large, exterior view windows overlook the primary hull. These windows can be opaqued if necessary. A small replicator provides suitable refreshments.

4.07.10 SANITARY FACILITIES

Two heads are located on the Main Bridge: one on the port side and one on the starboard side. Inclusion of these facilities enables Bridge watchstanders to minimize the amount of time away from stations during Bridge shifts.

4.07.11 LIFEBOATS

Six two-person lifeboats ring the Bridge. Launch of a lifeboat automatically activates a self-sealing barrier door which preserves Main Bridge atmospheric integrity.

4.08 MAIN ENGINEERING

Main Engineering, also known as the "Engine Room," is located in the secondary hull forward of the Shuttlebay and Cargo Hold 1. Within this complex are Propulsion Control, Environmental Control, Auxiliary Control, and a number of smaller compartments with specialized uses.

Engineering personnel monitor and control propulsion, power, power distribution, environmental conditions, thruster units, impulse engines, warp engines, and the matter/antimatter reaction unit from numerous computer consoles and data readout stations.

A central computer control panel in Main Engineering is the area where the Engineering Officer of the Watch (EOOW) stands duty. Critical readouts from all Engineering stations are repeated, allowing the EOOW to monitor gross functioning of the department. The functions of any other stations can be assumed from this console as necessary or desired.

4.08.01 PROPULSION CONTROL

This multi-purpose station monitors the status and functions of the three modes of propulsion available on the Starcruiser: thruster units, impulse engines, and warp drive. It is made up of a main section and two secondary sections.

The main section readouts provide a complete status of all components and energy outputs for the currently selected propulsion system. The two secondary sections provide basic data on the two systems not in use. Should the operator require additional information, the focus of the primary and secondary readouts can be interchanged.

Direct control over the propulsion systems can be rerouted at any time to personnel on watch. Emergency shutdown of any propulsion system, as well as jettison procedures, can be effected.

All functions can be transferred to the Engineering Console on the Main Bridge if directed by the Officer-of-the-Deck, Commanding Officer, First Officer, Chief Engineer, or Engineering Officer on Watch.

4.08.02 ENVIRONMENTAL CONTROL

Environmental Control monitors the performance of all life-support systems and provides a continual readout of the efficiency and effectiveness of artificial gravity, inertial dampening systems, atmosphere, and humidity.

Should a component fail, this console will automatically transfer the affected areas of the ship to another component. Depending upon the severity of the failure, maintenance is either begun immediately, or scheduled for a later time. Individual gravity generators can be altered to provide differing levels based on emergent requirements. Sections of the ship which are not used, do not require life-support (such as some sealed cargo bays), or are damaged can be isolated and life-support levels altered as necessary or desired.

4.08.03 AUXILIARY CONTROL

Four main systems are constantly monitored by Auxiliary Control:

Power Distribution
Replicators
Transporters
Closed-System Recovery Units

The main display provides a holographic view of the ship's interior, with each monitored system and/or component outlined in a different color. Power distribution circuits are shown in blue, replicators are in purple, transporters are orange, and the closed-system recovery matrix is yellow-green. Numerous sensors located at critical junctions or output nodes continually send status reports to Auxiliary Control. These status reports are compared by the console to established design and performance benchmarks.

If the sensor reading does not conform to these parameters, the appropriate section(s) of the display immediately change to flashing yellow. The operator can choose to enlarge the affected area. When the "zoom mode" is selected, both the expected and the actual status is displayed alpha-numerically. Maintenance or repair actions can then be initiated.

Components or circuits which meet specifications but are currently unused, are indicated by the color green. Red indicates damaged or out-of-commission items.

4.09 SCIENCE LABORATORIES

The extended hull of the AVENGER-class Heavy Frigate was utilized for Engineering spaces and shuttlebays. In the GALILEO-class Starcruiser, the extended hull is reserved almost exclusively for Science Department facilities. The Chief Science Officer's departmental office is also located here. Approximately one-fourth of the total available space is allocated each division within the Science Department.

Two computer systems and their associated spaces are housed in the extended hull: the Main Library Computer and the Science Department computer. Functions of either system can be assumed by the other should the necessity arise.

4.09.01 LIFE SCIENCES

Personnel assigned to the Life Sciences Division work in several major laboratories in the extended hull. Computer terminals, work space, desks, and other required equipment, specific to the particular area under investigation, are included in each compartment.

The Chief Life Sciences Officer maintains a divisional office in this area. Separate laboratories within the Life Sciences section of the Science Department space are designed to place associated specialists in close proximity to facilitate the interchange of knowledge and techniques.

Although separate laboratories are indicated, extensive interactions among the specialists in each facility is common and, in fact, a necessity. Therefore, these assignments are to be considered flexible.

These major spaces, and the specialists assigned that area, are:

Laboratory	Specialists
Animal	Biologist, Microbiologist, Geneticist, Ornithologist, Entomologist, Mammalologist, Herpetologist,

	Ichthyologist, Paleontologist, Marine Biologist, Zoologist
Plant	Agronomist, Botanist, Horticulturalist
Technologies	Bioengineer, Biotechnologist, Bio chemist, Cyrogenics Officer, Ecologist
Support Services	Chief Life Sciences Officer, Computer Analyst, Yeoman

4.09.02 PHYSICAL SCIENCES

Personnel assigned to the Physical Sciences Division work in several major laboratories in the extended hull. Computer terminals, work space, desks, and other required equipment, specific to the particular area under investigation, are included in each compartment.

The Chief Physical Sciences Officer maintains a divisional office in this area. Separate laboratories within the Physical Sciences section of the Science Department space are designed to place associated specialists in close proximity to facilitate the interchange of knowledge and techniques.

Although separate laboratories are indicated, extensive interactions among the specialists in each facility is common and, in fact, a necessity. Therefore, these assignments are to be considered flexible.

These major spaces, and the specialists assigned that area, are:

Laboratory	Specialists
Atmospheric	Cartographer, Climatologist, Geodesist, Physical Meteorologist, Planetologist, Synoptic Meteorologist
Chemistry	Analytical Chemist, Chemical Oceanographer, Geochemist, Inorganic Chemist, Organic Chemist

Geology	Economic Geologist, Exploration Physicist, Geochronologist, Geo magnetician, Geomorphologist, Mineralogist, Paleomagnetician, Paleontologist, Palynologist, Seismologist, Soil Scientist, Stratigrapher, Volcanologist
Hydrology	Geological Oceanographer, Hydrologist, Physical Oceanographer
Support Services	Chief Physical Sciences Officer, Chief Computer Systems Analyst, Yeoman, Computer Analyst

4.09.03 SOCIAL SCIENCES

Personnel assigned to the Social Sciences Division work in several major laboratories in the extended hull. Computer terminals, work space, desks, and other required equipment, specific to the particular area under investigation, are included in each compartment.

The Chief Social Sciences Officer maintains a divisional office in this area. Separate laboratories within the Social Sciences section of the Science Department space are designed to place associated specialists in close proximity to facilitate the interchange of knowledge and techniques.

Although separate laboratories are indicated, extensive interactions among the specialists in each facility is common and, in fact, a necessity. Therefore, these assignments are to be considered flexible.

These major spaces, and the specialists assigned that area, are:

Laboratory	Specialists
Contemporary	Curator, Demographic Researcher, Ethnoscientist, Geographer, Legal Specialist, Political Scientist, Sociologist
Historical	Anthropologist, Archeologist

| Support Services | Chief Social Sciences Officer, Yeoman, Computer Analyst |

4.09.04 SPACE SCIENCES

Personnel assigned to the Space Sciences Division work in several major laboratories in the extended hull. Computer terminals, work space, desks, and other required equipment, specific to the particular area under investigation, are included in each compartment.

The Chief Space Sciences Officer maintains a divisional office in this area. Separate laboratories within the Space Sciences section of the Science Department space are designed to place associated specialists in close proximity to facilitate the interchange of knowledge and techniques.

Although separate laboratories are indicated, extensive interactions among the specialists in each facility is common and, in fact, a necessity. Therefore, these assignments are to be considered flexible. These major spaces, and the specialists assigned that area, are:

Laboratory	Specialists
Astronomical	Astronomer, Astronomical Cartographer, Cosmologist, Pulsar Specialist, Quasar Specialist
Physics	Astrophysicist, Elementary Particle Physicist, Molecular Physicist, Nuclear Physicist, Plasma Physicist, Thermodynamic Physicist
Theoretical	Aeronautical Specialist, Aerospace Specialist, Astronautical Specialist
Support Services	Chief Space Sciences Officer, Computer Analyst, Yeoman

4.10 SICKBAY FACILITIES

The main Sickbay facilities are located together on the same deck. This arrangement facilitates the efficient operations of one of the most important areas of the ship. Patients can be examined, tests made, surgery performed, and recuperation accomplished within a relatively small area. Medical personnel are able to perform multiple, simultaneous tasks because of this arrangement and, therefore, fewer crewmembers are needed to perform the same level of tasking.

4.10.01 MAIN EXAMINATION CENTER

The Main Examination Center, located on the periphery of the complex in close proximity to turbolifts, is the first place a crewmember reports if in need of medical assistance. Initial examinations and classification of the injury or disease process are accomplished here. Once the patient is diagnosed, he or she can be directed to the appropriate area for treatment. Minor problems can be resolved by the examining physician or technologist. All other areas of Main Sickbay are accessible from the Main Examination Center.

4.10.02 CHIEF MEDICAL OFFICER'S OFFICE

Located adjacent to the Examination Center, the Chief Medical Officer's Office provides the Chief Medical Officer (CMO) with the privacy required for consultations, meetings, and other appropriate activities. From the CMO's computer terminal, diagnostics results, patient records and histories, the current life-signs readouts of any Sickbay bed, and the Medical database stored in the Main Computer Library can be accessed.

The CMO's Office has two main area: a conference room and a central area.

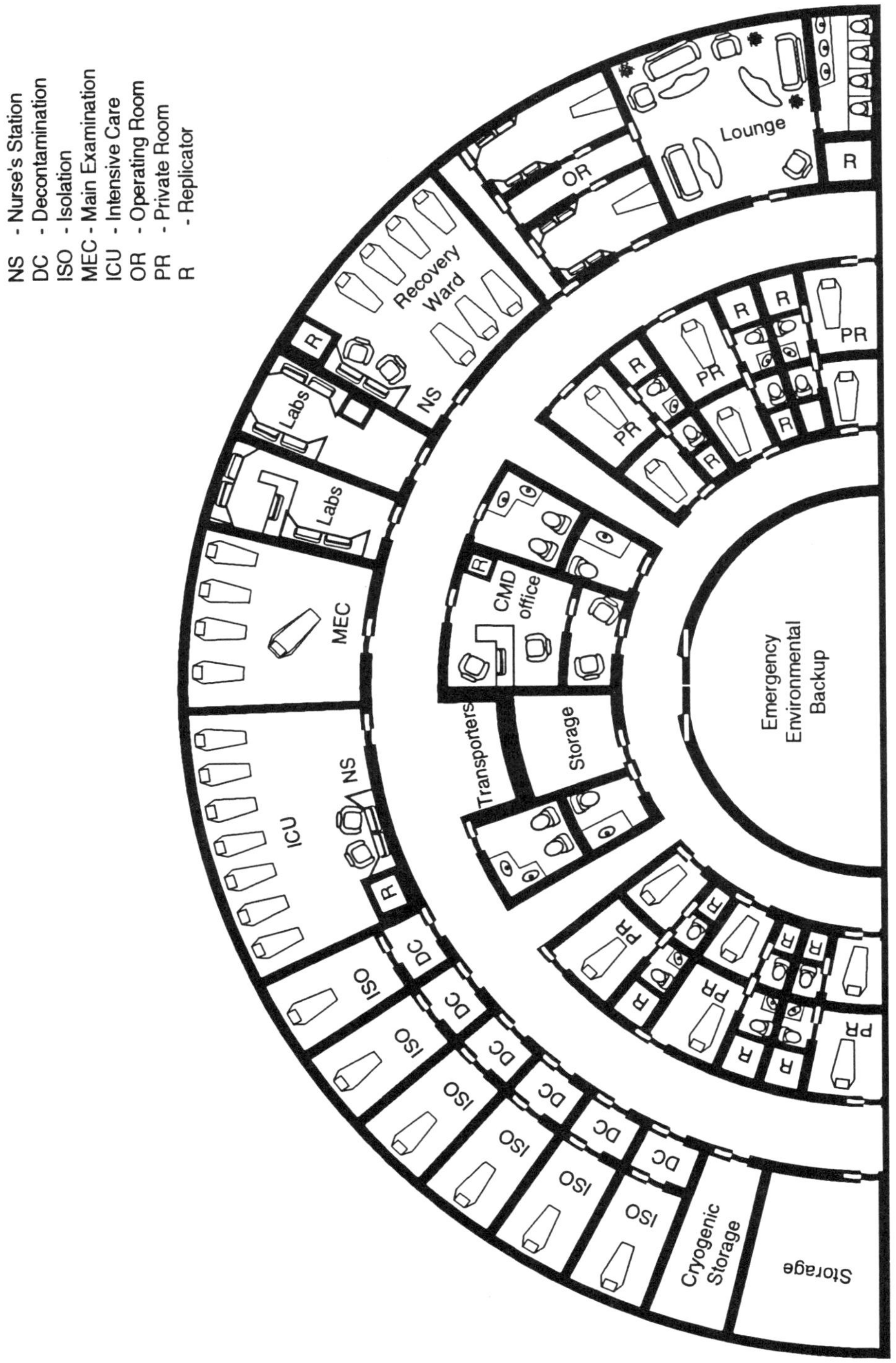

NS - Nurse's Station
DC - Decontamination
ISO - Isolation
MEC - Main Examination
ICU - Intensive Care
OR - Operating Room
PR - Private Room
R - Replicator

Lounge

OR

Recovery Ward

NS

Labs

Labs

MEC

ICU

NS

R

ISO
ISO
ISO
ISO
ISO
ISO

DC
DC
DC
DC

Cryogenic Storage

Storage

PR
PR
PR

R
R
R
R

CMD office

Transporters

Storage

Emergency Environmental Backup

PR
PR

R
R

Main Sickbay Complex

4.10.03 SECONDARY SICKBAY

A Secondary Sickbay is located just beneath the forward bulkhead of the Shuttlebay. This complex contains a main examination area and a 22-bed ward, both of which are scaled-down versions of Main Sickbay facilities. It is utilized for overflow patients and during mass casualty situations for patients with minor injuries so that Main Sickbay can handle more serious cases. During normal operations, Secondary Sickbay is not manned by Medical personnel, although all equipment and supplies are inspected on a regular basis.

4.10.04 WARDS

As in any modern medical complex, patients have differing needs. Some require around-the-clock observation and care, while others have lesser requirements. Three different wards are provided to handle this variety of care levels.

4.10.04.01 ISOLATION

In some cases, a patient may display symptoms of a disease which is highly contagious. These diseases may be airborne or spread by contact. To prevent infection of the remainder of the crew or other patients, an Isolation Ward segregates these types of patients. Designed to handle up to six patients with completely different disease manifestations, each isolation chamber can be programmed for a wide variety of atmospheric and gravity requirements should the particular disease warrant a change from normal shipboard conditions.

No one can enter an isolation chamber in the ward without the appropriate code and authorized BSC entered into the chamber door control panel. An attempt at unauthorized entrance causes the ward to be sealed and an alarm to be sounded in both the CMO's Office and the Security Duty Office.

4.10.04.02 Critical/Intensive Care

Seriously injured personnel and those requiring close monitoring of disease processes, are placed into the Critical/Intensive Care Unit. Biobeds for seven patients and a nursing station are in this compartment.

4.10.04.03 Recovery

Up to seven patients can utilize the Recovery Ward simultaneously. It is used as a post-operative staging room where the biomedical status of patients can be closely monitored. Should circumstances dictate, a patient may remain in this Ward for several hours. Any patient who is kept for observation (overnight, for example) will remain in this Ward.

Patients who require longer term recuperation are placed into one of the 12 private rooms.

4.11 Shuttlecraft Facilities

GALILEO-class Starcruisers carry a larger number of shuttlecraft than a normal starship because of the ship's primary mission. As a result, the facilities required to launch, recover, store, and perform maintenance and repairs are considerably larger. The shuttlecraft complex, composed of the Shuttlebay, Flight Deck Control, and the Repair and Maintenance Center, are located in the secondary hull.

4.11.01 Shuttlebay

The Shuttlebay is primarily used to launch and recover shuttlecraft. However, it can also be used for large gatherings, such as personal awards ceremonies, personnel inspections, and change-of-command ceremonies.

4.11.02 Flight Deck Control

The Pilot-in-Command (PIC) of a shuttlecraft has complete authority over the craft and its occupants once launched from the Shuttlebay. However, this authority is transferred to the Flight Deck Control Officer when the craft passes through the Shuttlebay doors.

The Flight Deck Control Officer has an unobstructed view of the Shuttlebay from Flight Deck Control, which takes up the complete forward bulkhead of the Shuttlebay. Atmosphere, gravity, the shuttlecraft conveyor/storage unit, and Shuttlebay doors are all controlled from this location. A tractor beam unit, mounted on the fantail, or extreme end of the flight deck, is used, when necessary, to recover immobile shuttlecraft or other, non-powered items.

4.11.03 REPAIR AND MAINTENANCE CENTER

Situated immediately below the Shuttlebay, the Repair and Maintenance Center is co-located with the shuttlecraft storage area. When not scheduled for use or in need of repair or maintenance, shuttlecraft are moved into a circular conveyor system accessed on either side of the Shuttlebay. Designed to accommodate a full-sized craft, the individual compartments of the conveyor/storage system move in a circular path down to the Repair and Maintenance area.

There are two distinct areas on the Repair and Maintenance Deck: minor repair and maintenance, and major overhaul; craft are placed in the appropriate area depending upon the expected or planned tasks to be performed. Variable gravity generators enable maintenance personnel to remove or replace large components with ease. Because of the dangers involved in this variable gravity area of the ship, it is accessible only by authorized and specially trained mechanics, pilots, and engineers.

4.12 RECREATIONAL FACILITIES

Long, stressful voyages, the norm for GALILEO-class Starcruisers, require that a wide variety of relaxation facilities be made available to the crew. Specialists in recreation, music, physical activities, and Thespian activities are assigned to assist in organizing recreational activities.

4.12.01 AUDITORIUMS

One small auditorium, designed for play production and ensemble performances, is in the Primary Hull. Two dressing rooms, one on each wing of the stage, are available for costume changes. Engineering Department personnel are available for fabrication of period costumes.

The Main Library Computer carries a wide variety of scripts. A crewmember who desires to produce his or her own stage play, should contact the Thespian Instructor in Logistics Department, Recreation Division, for assistance. A main auditorium for large gatherings is located in the secondary hull.

4.12.02 BOTANICAL GARDEN

Designed to provide a relaxing change of scenery for nature lovers, the Botanical Garden replicates the environment and ecology of a Class-M planet. The wandering paths and cul-de-sacs create pockets of privacy. Numerous plants native to dozens of United Federation of Planets member worlds create a wonderland of botanic delights to help alleviate the stress incumbent to star ship duty.

The grounds are maintained by volunteers and the horticulturist assigned to Life Sciences Division of the Science Department. No restrictions exist for use of the Botanical Garden; they are available to any crew member or visiting VIP at any time of the day or night. Additions to the flora are welcome; however, before bringing any living plant or animal aboard the ship, Science Department personnel must be advised.

4.12.03 GYMNASIUMS

The often sedentary lifestyle on a Starcruiser and the need to maintain optimum physical fitness requires suitable facilities for individual workouts as well as team or individual sports. Three different types of gymnasiums aboard the ship provide appropriate space for such activities: a Standard Gravity Gym, a Zero-Gravity Gym, and a High-Gravity Gym.

4.12.03.01 STANDARD GRAVITY

Available to all personnel for various activities ranging from individual gymnastics to team sports such as volleyball, basketball, rollerball and Pins, gravity in this gymnasium is maintained at one Earth-standard level at all times. Another Standard Gravity Gymnasium is located in the secondary hull.

Schedules for team sports and reservations for individual sports may be made by contacting the Gymnasium Facilities Officer.

4.12.03.02 ZERO GRAVITY

Almost directly opposite the Standard Gravity Gymnasium is a smaller facility designed for use in zero or near-zero gravity. Controls located within the room allow for gravity changes from one Earth Standard gravity down to zero gravity. Another Zero Gravity Gymnasium is located in the secondary hull.

4.12.03.03 HIGH GRAVITY

Next to the Zero Gravity Gymnasium is a gymnasium designed for use at above Standard Earth gravity. Controls located within the room allow for gravity changes above one Earth Standard, up to a maximum of 10 Earth Standard gravities. Another High Gravity Gymnasium is located in the secondary hull.

4.12.04 LOUNGES

A number of lounges are available on board the ship. Most are open to all members of the crew at any time. Other lounges are restricted in use, such as the VIP Lounges. Crewmembers must ensure they do not attempt to utilize a lounge which may be restricted.

4.12.04.01 MAIN OBSERVATION LOUNGE

The Main Observation Lounge is an unrestricted lounge designed for personnel of all ranks to share. Seating is available for approximately 80 off-duty personnel. The Main Observation Lounge is manned by at least one Logistics Department server at all times.

Six large viewing ports provide an unobstructed view of space and the primary hull. A fountain on each end of the Main Observation Lounge contribute to the relaxed atmosphere. Food processors on the inboard bulkhead offer a wide selection of alcoholic and non-alcoholic beverages as well as appetizers or standard meal fare.

4.12.04.02 SECONDARY LOUNGES

Both the primary hull and the secondary hull have several secondary, unmanned lounges at various locations. Each one has a food processor identical to the ones located in the Main Observation Lounge. Some of these secondary lounges are positioned so as to have exterior views; others do not. Secondary lounges are unrestricted.

4.12.04.03 VIP LOUNGES

Two lounges are reserved for VIPs use only. Both are located on the periphery of the hull so that external views are available. These lounges are not staffed unless the ship is carrying guests eligible to use them. At that time, depending upon the number of VIPs aboard, one or more staff are assigned duty in VIP Lounges to serve as necessary.

One VIP Lounge is in the primary hull and the other in the dorsal. Small receptions are often held in one of these lounges (or, in some cases, in the Captain's Suite). Large receptions are usually held in the Main Observation Lounge.

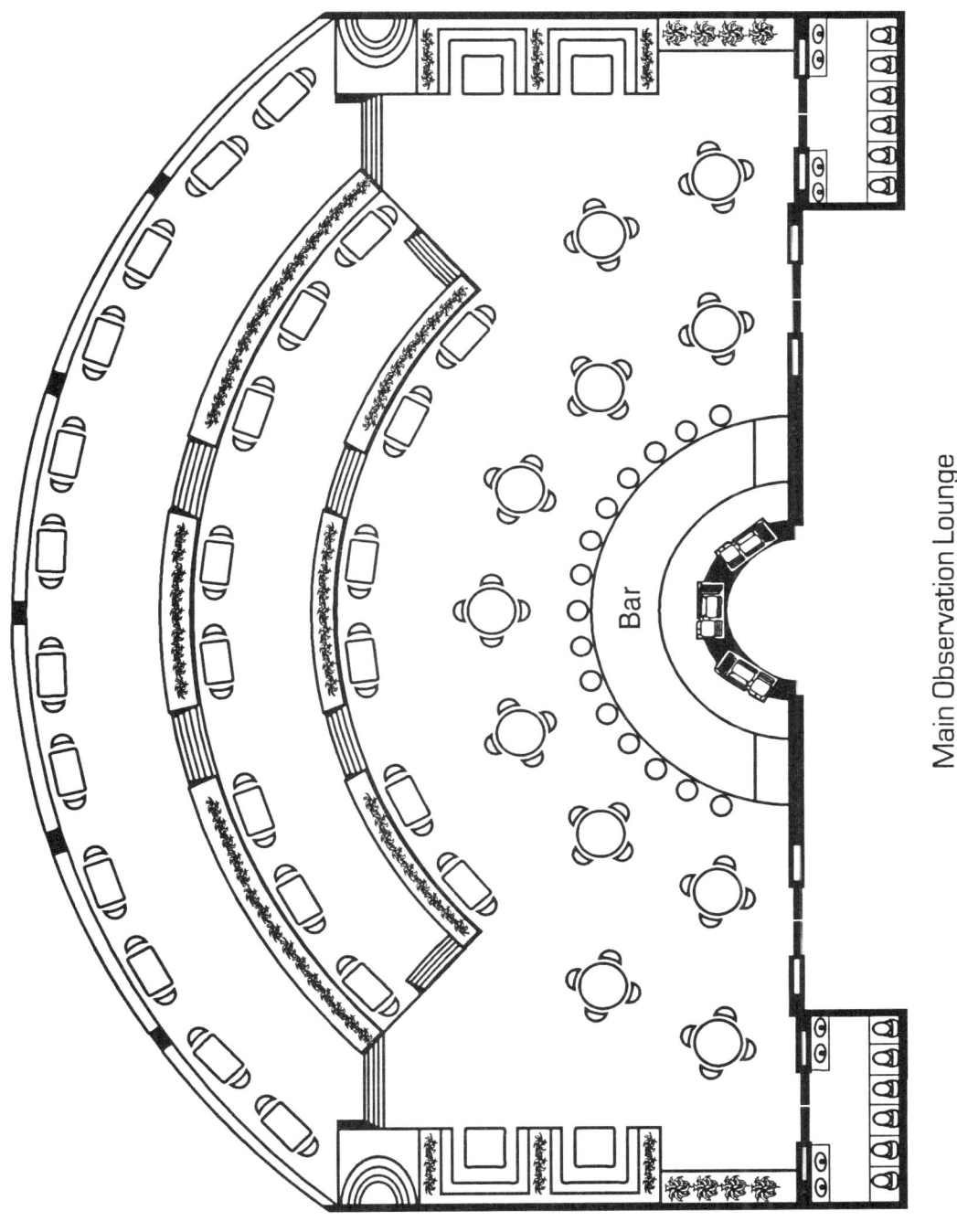

Main Observation Lounge

Bar

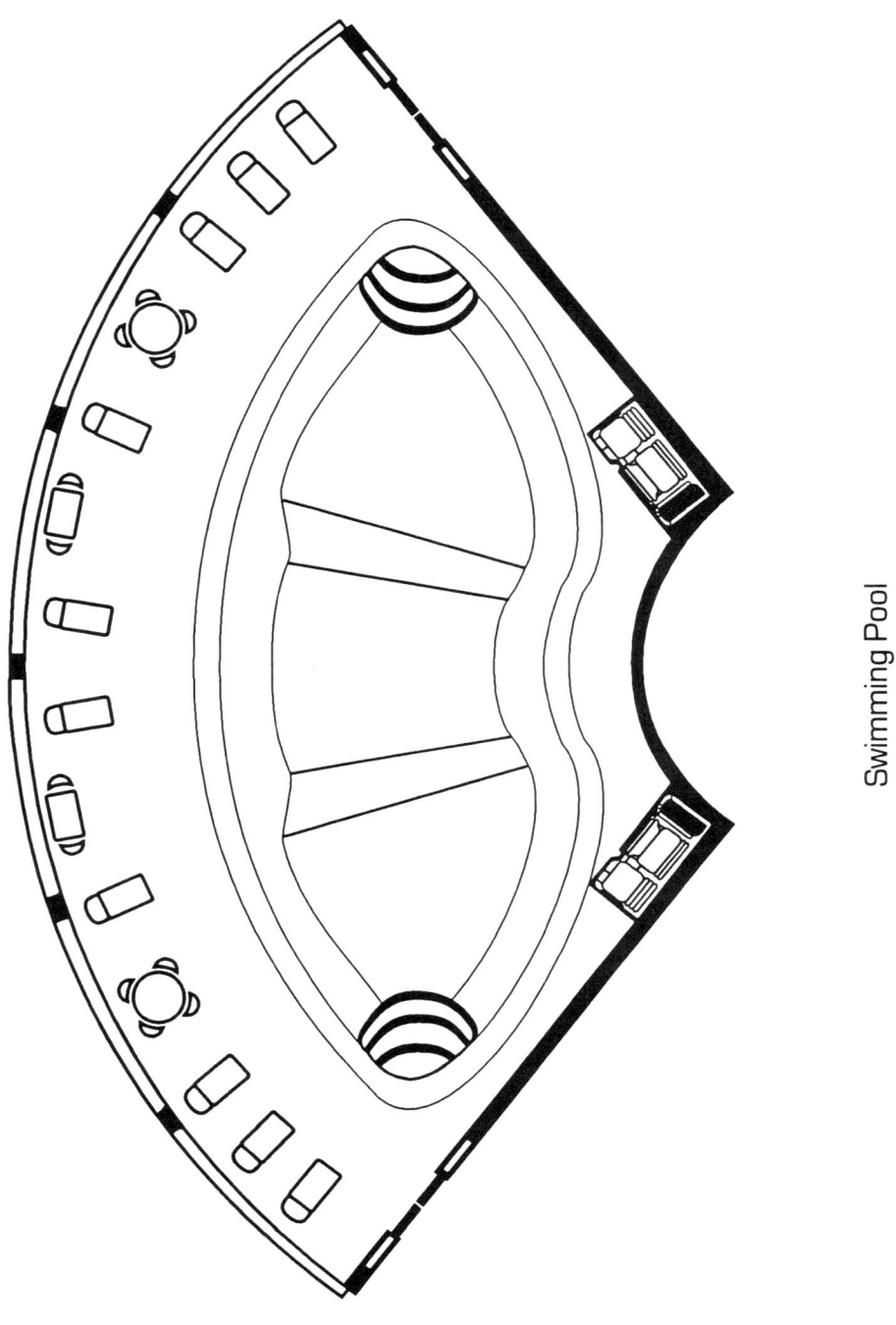

Swimming Pool

4.12.05 GAME ROOMS

Also located throughout the ship are game rooms of several varieties, each of which provides a combination of Computer Games (such as Tactical Simulation, Chess, StarTraxs: The Final Gamble, Risk, and others) and Table Games (such as Pool, Table Tennis and Pook-Kai). Game rooms are open continuously and no restrictions are placed on their use.

4.12.06 SWIMMING POOL

Depths range from three feet at each end to a maximum depth of 10 feet at the center. Ample space is provided around the pool for relaxation and consumption of beverages. Six viewing ports provide exterior views.

4.13 SENSOR PACKAGES

Sensors are classified as either long-, medium-, or short-range, depending upon the system's maximum delivery radius. They are further categorized as either passive or active, based upon the specific sensor methodology utilized. Two main types of sensor packages are used aboard Starcruisers: sensor arrays and probes. Each is discussed below. Tricorders, which fall into the sensory array type, will not be addressed.

4.13.01 LATERAL SENSOR ARRAY

The newest in sensor design, the GALILEO-class Starcruiser utilizes a system which is integral to the primary hull. The "OMNI-Synse" is contained in the circumference of the primary hull and, as a result of both the hardware and software design of the system, extends the effective sensor range by 125%, and improves the resolution capabilities by 150% over the standard sensor suite. Because of this unique design, the sensor package normally located at the bottom of the primary hull has been removed.

4.13.02 PROBES

Nine different classes of probes are carried aboard. Each is designed for a specific task, although modifications can be made which will allow the mission of one class to be accomplished by another class.

4.13.02.01 CLASS I
(Sensor Probe—FIR-44-1)

This is the standard, short- to medium-range sensor package, which is most often utilized for general situations. Its 200,000 km range enables the ship to remain at a safe distance from the object or anomaly to be investigated.

The payload of a Class I probe includes a standard sensor package; an electromagnetic resolution unit; and an extensive diagnostic chemistry laboratory, which can accomplish detailed subspace and interstellar chemical analysis.

A total of 20 Class I probes are carried on board the GALILEO-class Starcruiser, of which 10 are maintained in a "ready" mode. The remainder are stored in a standby status.

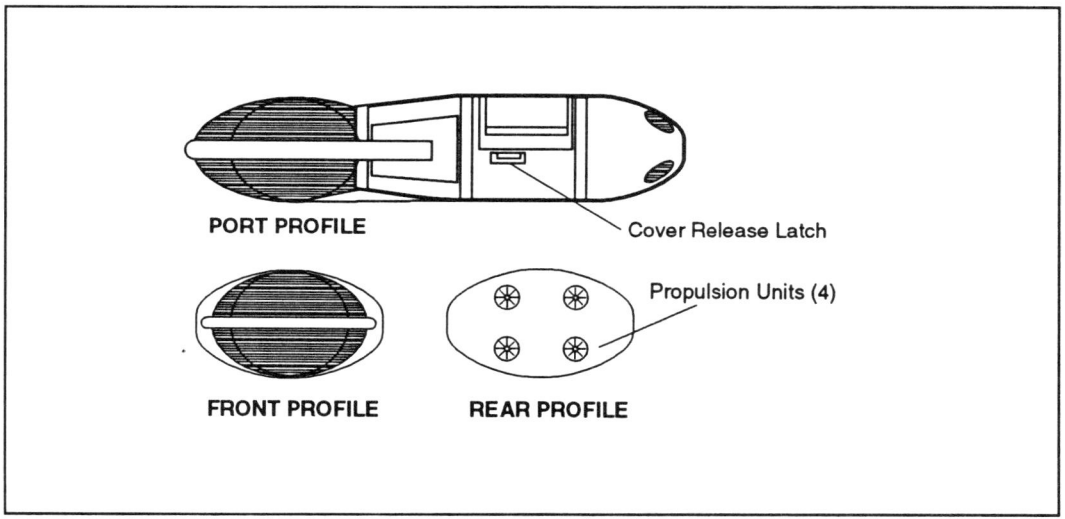

DFIR-44-1 Sensor Probe

4.13.02.02 CLASS II
(Sensor Probe—SENPROB GSU-2-2)

This long-range probe has twice the effective radius of the Class I. It carries the same standard sensor package, electromagnetic resolution unit, and diagnostic chemistry laboratory as the Class I. In addition, long-range particle detectors and field detectors are included in the normal payload. An imaging system can be included should it be necessary.

A total of 25 Class II probes are carried on board the GALILEO-class Star cruiser, 10 of which are maintained in a "ready" mode. The remainder are stored in a standby status.

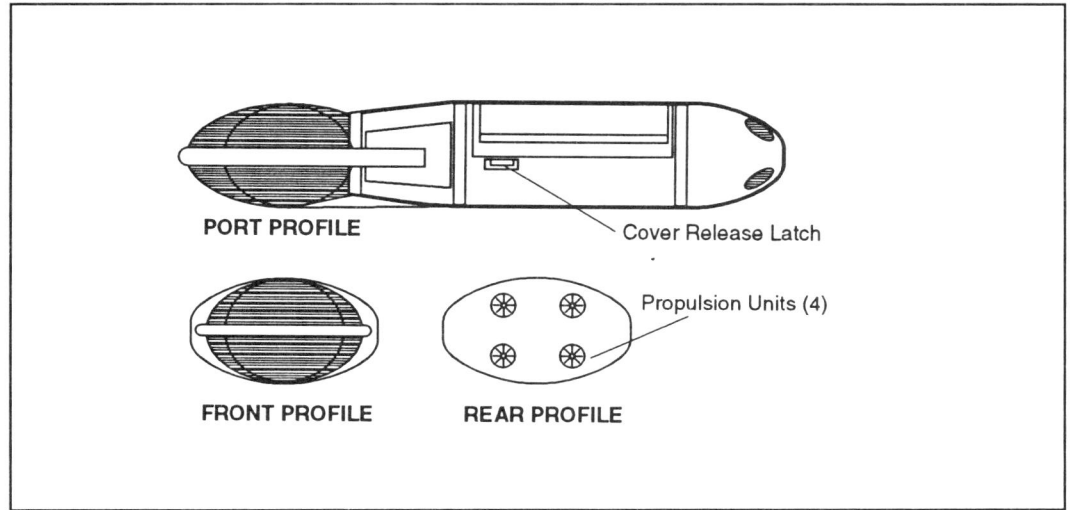

SENPROB GSU-2-2 Sensor Probe

4.13.02.03 CLASS III
(Planetary Probe—PLNPROB QMJ-12-3)

A 450-bar pressure reinforced hull allows this probe to soft land on any planet with less than a 3-gravity well. The usual payload includes the standard sensor package, field detectors, gas giant and planetary sensing units. Also available for inclusion are an on-board chemical analyzer and material sampler. The Planetary Probe has the ability to remain in an assigned area for an extended period of time, providing a measure of continuity in observations.

A total of 20 Class III probes are carried on board the GALILEO-class Star cruiser, of which 10 are maintained in a "ready" mode. The remainder are stored in a standby status.

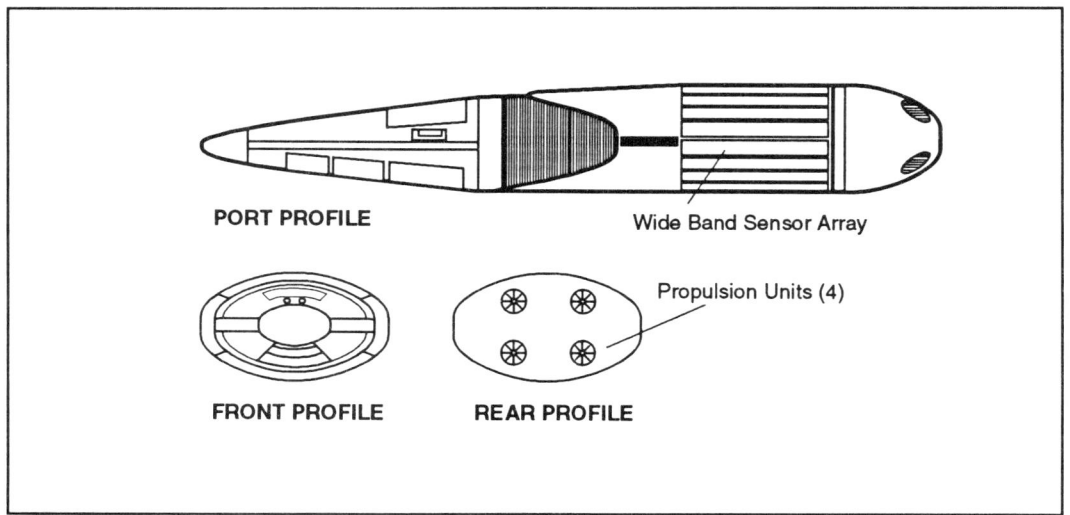

PLNPROB QMJ-12-3 Planetary Probe

4.13.02.04 CLASS IV
(Stellar Encounter Probe—STLPROB HHT-55-4)

Temperatures close to a stellar body, and in many cases stellar remnants, can overwhelm a normal sensor probe. To alleviate this, the Stellar Encounter Probe employs not only a special deflector shield system, but also extensive heat sinks for dissipation. Included with the standard sensor package are stellar field sensors, particle detectors, and stellar atmosphere detectors. Six deployable radiation flux sensor units expand this probe's capabilities.

A total of five Class IV probes are carried on board the GALILEO-class Starcruiser, one of which is maintained in a "ready" mode. The remainder are stored in a standby status.

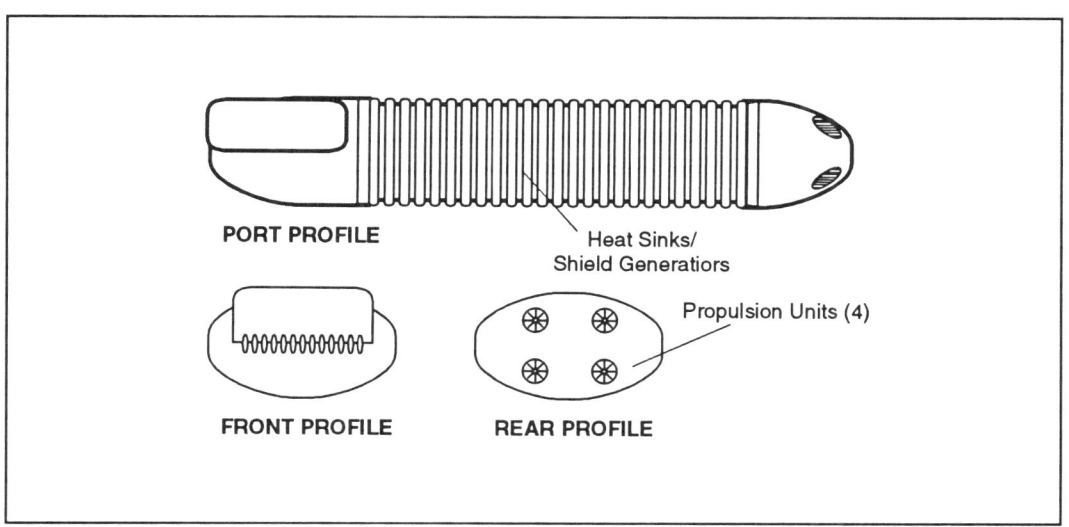

STLPROB HHT-55-4 Stellar Encounter Probe

4.13.02.05 CLASS VI

(Communications Relay/Emergency Beacon— COMPROB DGY-71-6)

This dual-purpose unit can be one of the most important items in a Star cruiser's probe inventory, even though the term "probe" may be considered a misnomer. When utilized as a communications relay, it boots the strength of signals to increase their effective range. Both starships and probes can use its facilities. It is often used to replace defective communication buoys. Its high-gain antenna, extended power supply, and expanded communication channels provide an ideal platform for an emergency beacon.

A total of five Class V probes are carried on board the GALILEO-class Starcruiser, one of which is maintained in a "ready" mode. The remainder are stored in a standby status.

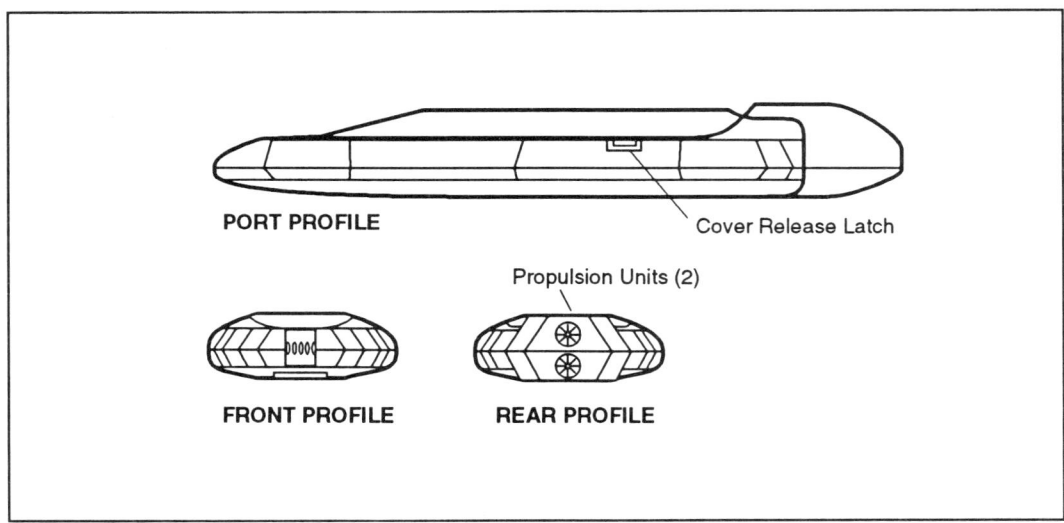

COMPROB DGY-71-6 Communications Relay/Emergency Beacon

4.13.02.06   Class VII
(Remote Culture Study Probe— CLTPROB TLU-1-7)

Expanded passive sensors, extremely low observability, a long station-keeping capability, and a planetary soft-landing ability contribute to preserving the Prime Directive. Should the probe be discovered by the culture under investigation, automatic procedures are instituted which reduce all electronic and mechanical components to basic elements, preventing cultural contamination.

A total of 10 Class VII probes are carried on board the GALILEO-class Starcruiser, two of which are maintained in a "ready" mode. The remainder are stored in a standby status.

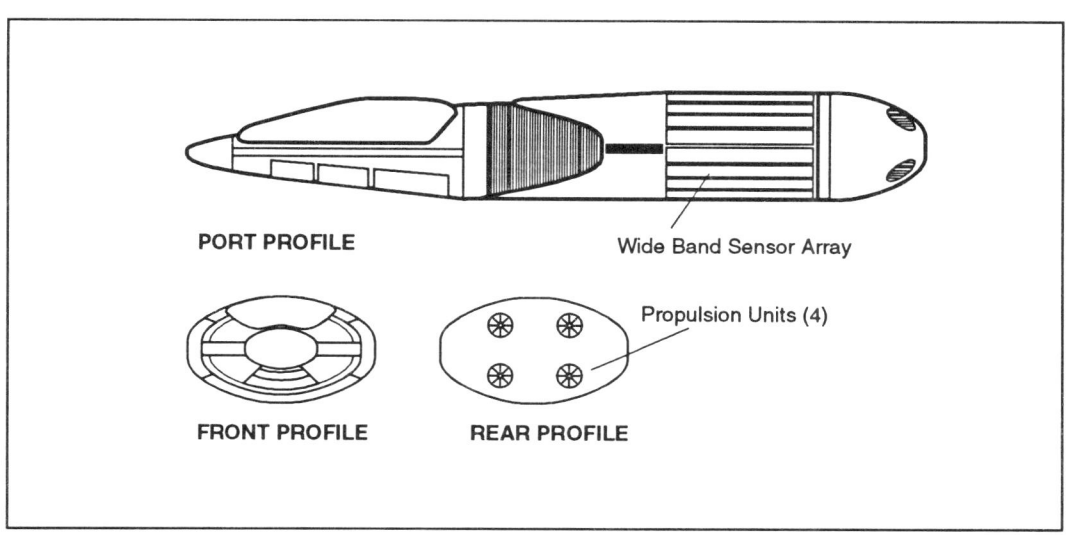

PORT PROFILE

Wide Band Sensor Array

FRONT PROFILE

REAR PROFILE

Propulsion Units (4)

CLTPROB TLU-1-7 Remote Culture Study Probe

4.13.02.07 Class VIII

(Medium-Range, Multipurpose Probe—WRWPROB DFC-45-8)

This probe can be configured with a number of different sensor modules to reflect a mission-specific role. Should the necessity arise, it is able to mimic the characteristics of any other class probe. It is most often used for special experiments which require tailor-made payloads.

A total of five Class VIII probes are carried on board the GALILEO-class Starcruiser, one of which is maintained in a "ready" mode. The remainder are stored in a standby status.

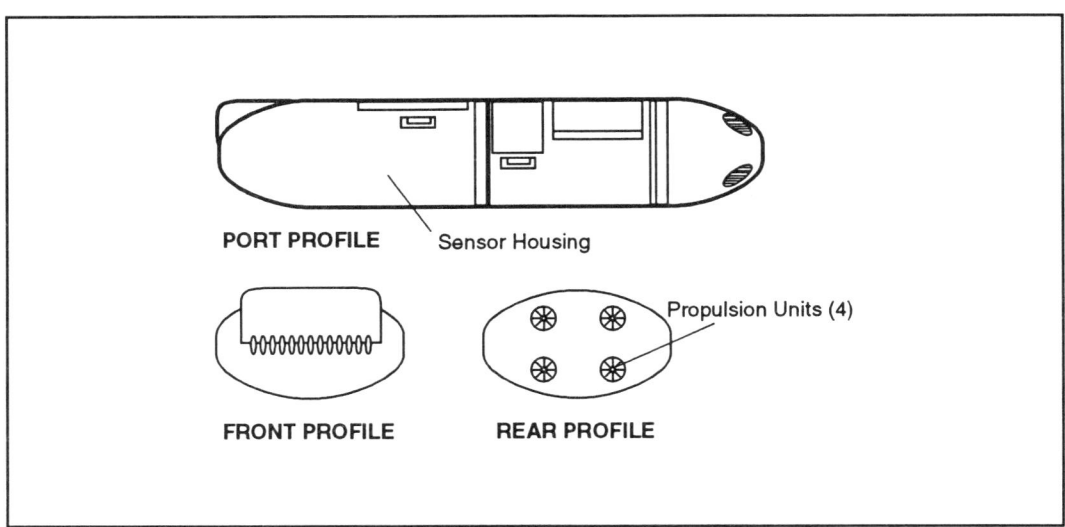

WRWPROB DFC-45-8 Medium-range, Multipurpose Probe

4.13.02.08 Class IX

(Long-range, Multipurpose Probe—LRWPROB SWS-B-9)

This probe is identical in concept to the DFC-45-8. The major difference is its flight radius, which is approximately 700% greater than the Class VIII probe.

A total of five Class IX probes are carried on board the GALILEO-class Starcruiser, one of which is maintained in a "ready" mode. The remainder are stored in a standby status.

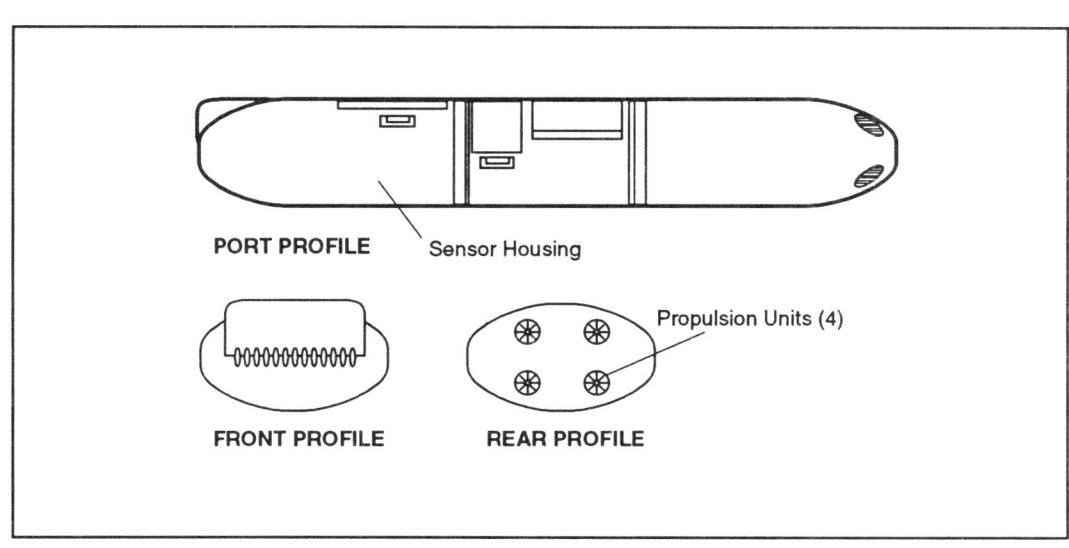

LRWPROB SWS-B-9 Long-range, Multipurpose Probe

4.14 SITUATION ROOM

The Situation Room is located adjacent to the Auxiliary Bridge in the Primary Hull. It is activated only when a Starcruiser functions as a Fleet Coordination Vessel or if directed by higher authority. It provides command, control, and communications facilities for a Task Force Commander to manage assigned assets during periods of hostilities or catastrophic contingencies.

When activated, the Situation Room is manned by personnel assigned to the Task Force Commander. Under normal circumstances, shipboard personnel are not utilized in the Situation Room. However, should the need arise, selected ship's company personnel are trained to staff the Situation Room until suitable replacements can be provided.

Typical Situation Room manning includes the following:

Position	Minimum Rank and Number
Task Force Commander	RADM (1)
Task Force Assistant	CAPT (1)

Position	Minimum Rank and Number
Intelligence Officer	CDR (3)
Watch Officer	LCDR (3)
Tactical Display Console Operator	LT (9)
Communications Console Operator	LT (9)
Ancillary Staff	As required by Task Force Commander

4.15 STATEROOMS

The Chief Administrative Officer will assign living quarters immediately upon an individual's reporting on board for duty. There are several different types of quarters and each is assigned based upon rank and quarters' availability.

Although berthing assignments, like duty assignments, are designed to be non-gender specific, assignment to quarters aboard the ship will be driven by appropriate cultural and social norms of the crew member. In addition, personality traits are considered to be of paramount importance. Every attempt will be made to assign compatible individuals to double-occupancy staterooms.

If, however, an individual should discover that a co-occupant is incompatible, the Chief Administrative Officer is to be notified immediately so that alternative arrangements can be investigated.

Should the type of quarters for which an individual is eligible not be available, an assignment will be made to one at a higher level. For instance, should the crewmember qualify for Level D quarters and all such quarters are filled, berthing assignment will be made to Level C quarters. As soon as the appropriate quarters are available, the individual will be reassigned.

If a promotion qualifies the crewmember to move from a double- to a single-occupancy stateroom, the option to remain in the current quarters is available.

Fabricators assigned to the Engineering Department are available to crew members who wish to personalize their quarters (within certain parameters).

4.15.01 COMMANDING OFFICER SUITE

The Commanding Officer's quarters, or Suite, is designed to reflect the stature of the senior officer aboard a major fleet unit. The Suite is composed of three majors sections: a briefing room, living/entertaining area, and sleeping/sanitary facilities.

The briefing room contains a conference table with places for the Commanding Officer, the First Officer, all department heads, and the MIMAD officer (when embarked). A food processor is available if the briefing room is used for formal dining. Two viewscreens are programmable from controls located by each chair. A small head is available for use.

A large living/entertaining area separates the briefing room from sleeping/sanitary facilities. A number of chairs and sofas provide for the comfort of guests and are arranged to facilitate pleasant conversation. A fountain and several planters contribute to the relaxed atmosphere.

The sleeping area also has a private computer alcove and library for research or pleasure. A raised spa, complete with a small food processor, is reached from the sleeping quarters via steps. Other amenities include a large, walk-in closet, and a Jacuzzi tub with a sonic/water-selectable shower.

All exterior doors, with the exception of the main corridor entrance to the briefing room, are voice-keyed to respond only to the Commanding Officer's voice, or other designated individuals.

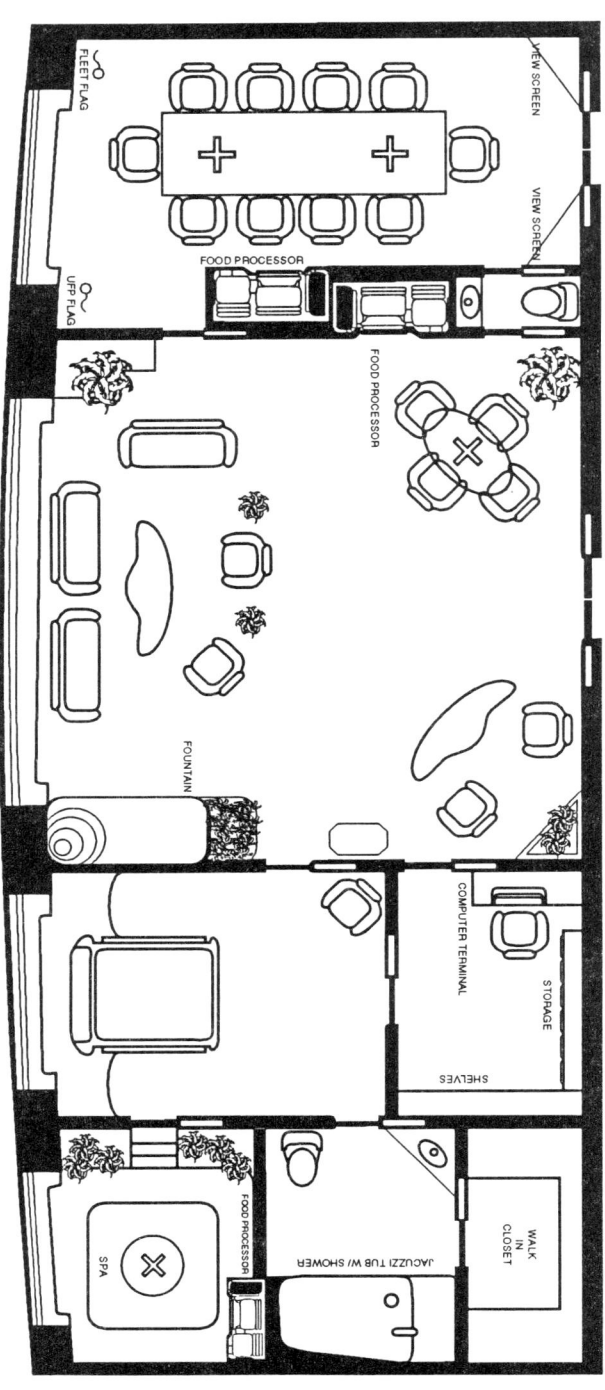

Commanding Officer's Suite

4.15.02 LEVEL A

These large suites are reserved for the First Officer and high-ranking visitors aboard the ship.

Each one contains a conference room, an entertainment area, a private study with computer facilities, a sleeping area, and sanitary facilities. Two food processors, one in the entertainment area and one in the conference room, provide for private dining. Head facilities that include a Jacuzzi tub and a sonic shower, are designed to accommodate two individuals.

Large view windows are located in the private study and the entertainment area. The conference room also contains a large viewscreen for use during briefings. Wide latitude is allowed the First Officer in decorating these quarters. Individual environmental controls allow for simulation of home world humidity, temperature, and light spectrum.

Twelve Level A quarters (in addition to the one reserved for the First Officer) are located in the primary hull, secondary hull and the dorsal for visitors who meet assignment criteria.

4.15.03 LEVEL B

There are two types of Level B quarters: B_1 and B_2. Level B_1 quarters are single-occupancy quarters for Department Heads and the MIMAD Commanding Officer (when embarked), containing a conference room, a living area, and a sleeping area. Two food processors, one in the living area and one in the conference room, provide for private dining. Head facilities include a Jacuzzi tub with a shower as well as a sonic shower.

Wide latitude is allowed in decorating these quarters. Individual environmental controls allow for simulation of home world humidity, temperature, and light spectrum. Eight Level B_1 quarters are located in the primary hull; five additional Level B_1 quarters are located throughout the ship for supercargo personnel who meet assignment criteria.

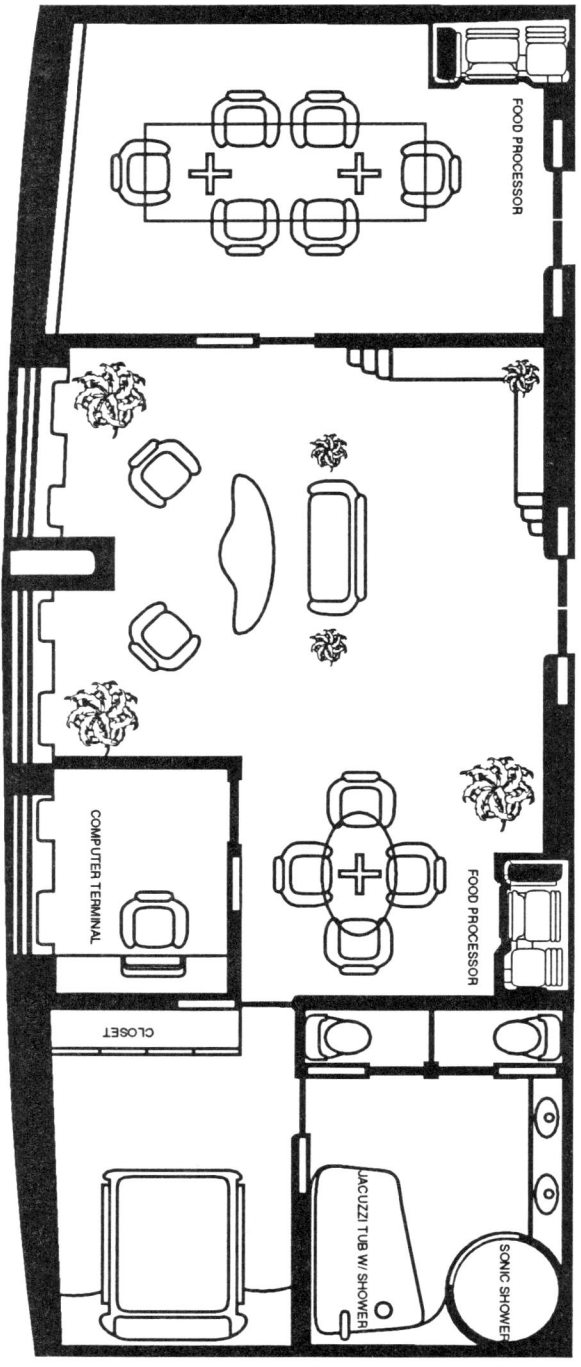

Level A Quarters (First Officer and VIPs)

Level B_2 quarters are single-occupancy quarters for Lieutenant Commanders and Commanders (or equivalent), containing a living area and a sleeping area. A food processor is located in the living area for private dining. Head facilities include a Jacuzzi tub with a shower as well as a sonic shower. Wide latitude is allowed permanently assigned ship's personnel in decorating these quarters. Individual environmental controls allow for simulation of home world humidity, temperature, and light spectrum.

Twenty-seven Level B_2 quarters are located throughout the ship; five additional Level B_2 quarters are located throughout the ship for supercargo personnel who meet assignment criteria.

4.15.04 LEVEL C

Double-occupancy quarters for junior officers (Ensign, Lieutenant (jg), and Lieutenant—or equivalent), with separate sleeping areas and shared living area and head.

Wide latitude is allowed permanently assigned ship's personnel in decorating these quarters. Individual environmental controls allow for simulation of home world humidity, temperature, and light spectrum.

A total of 160 Level C quarters are located throughout the ship with an additional 10 available for supercargo personnel who meet assignment criteria.

4.15.05 LEVEL D

Single-occupancy quarters for senior enlisted personnel (Chief Petty Officer and above—or equivalent).

Wide latitude is allowed permanently assigned ship's personnel in decorating these quarters. Individual environmental controls allow for simulation of home world humidity, temperature, and light spectrum.

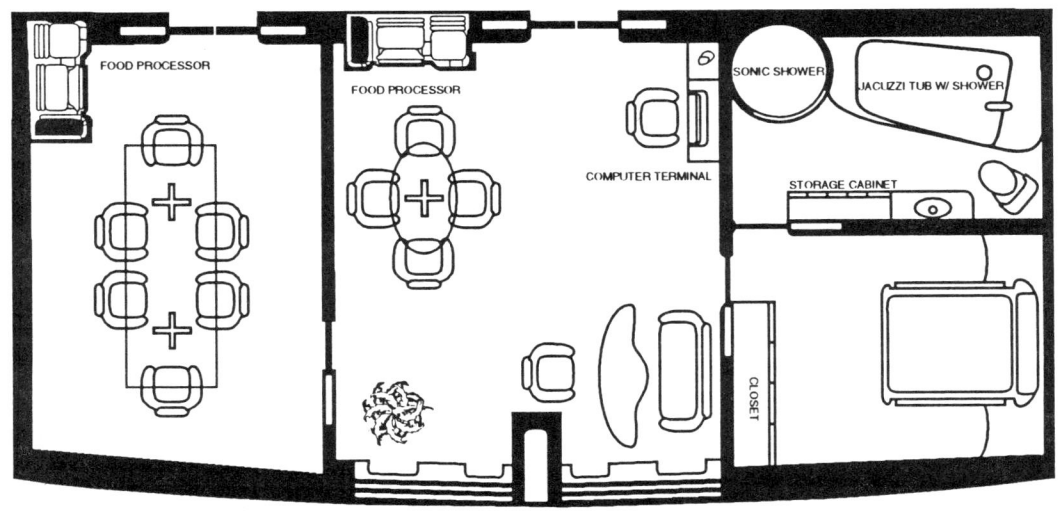

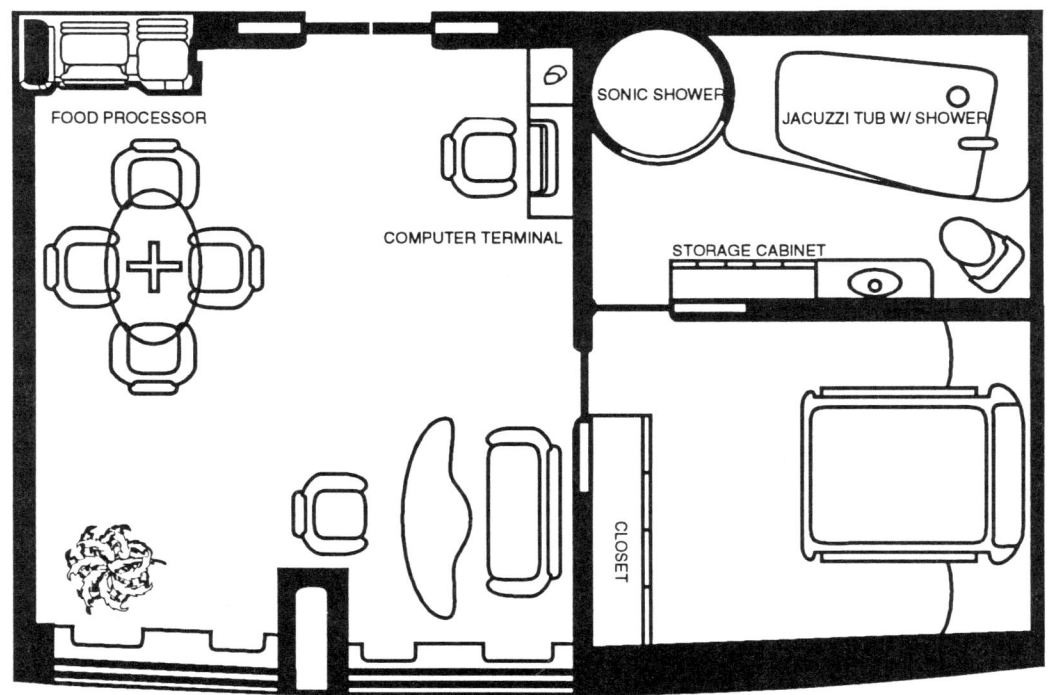

Level B₁ and Level B₂ Quarters

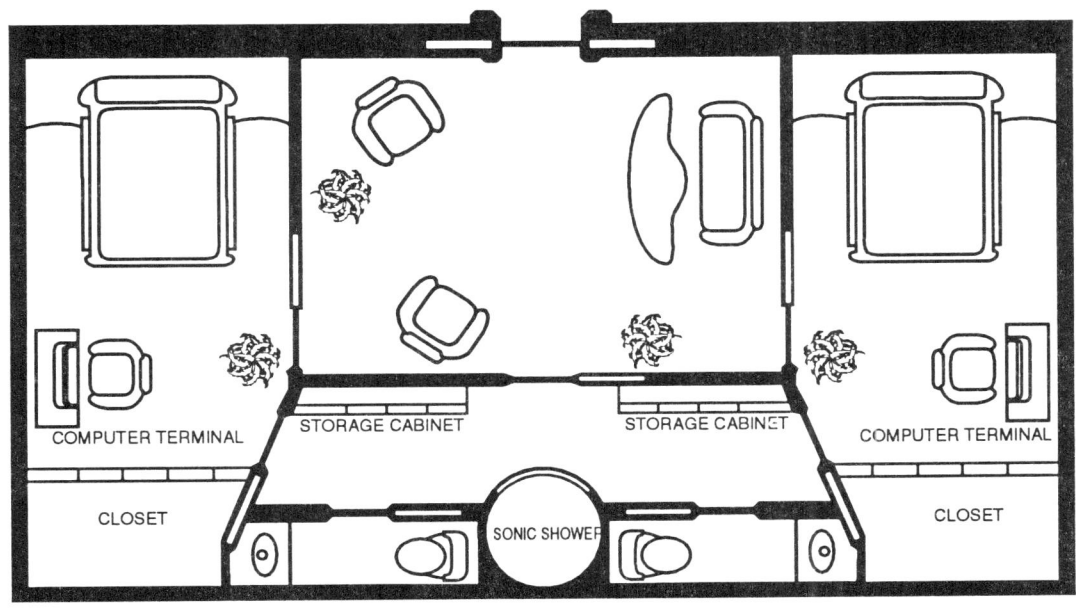

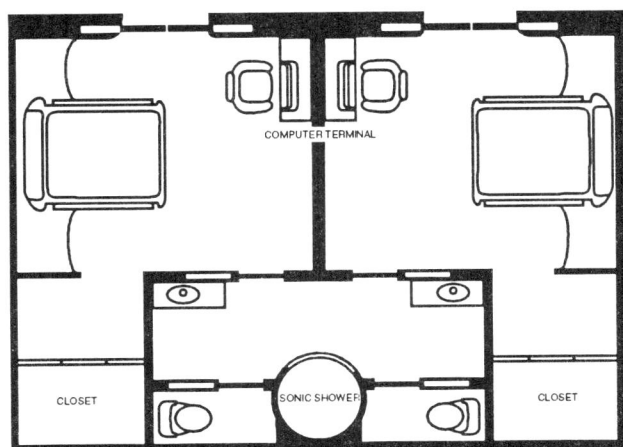
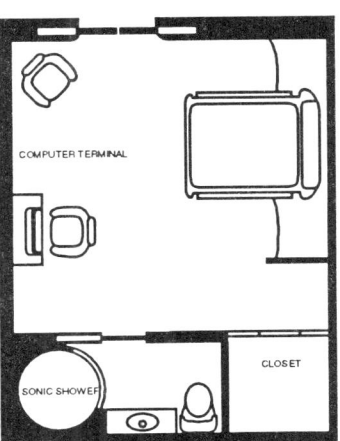

Level C, Level D, and Level E Quarters

A total of 35 Level D quarters are located throughout the ship with an additional 10 available for supercargo personnel who meet assignment criteria.

4.15.06 LEVEL E

Double-occupancy quarters for junior enlisted personnel (Petty Officer Third Class to Petty Officer First Class—or equivalent) with shared sanitary facilities and separate sleeping areas.

Wide latitude is allowed permanently assigned ship's personnel in decorating these quarters. Individual environmental controls allow for simulation of home world humidity, temperature, and light spectrum.

A total of 104 Level E quarters are located throughout the ship with an additional 10 available for supercargo personnel who meet assignment criteria.

4.16 TRANSPORTERS

Three different types of transporters are used on the GALILEO-class Starcruiser: personnel, medical, and cargo. While the operation of each is identical, minor differences in design and usage are pointed out below.

4.16.01 PERSONNEL

There are two kinds of personnel transporters: standard and emergency. A total of six standard, six-person units, and four emergency, 23-person units, are located throughout the ship.

Three standard transporter complexes are located in the primary hull, and three in the secondary hull. These high-resolution, organic-capable units are used to transport and receive individuals over distances up to 30,000 kilometers.

The emergency evacuation units can handle up to 23 individuals simultaneously. Because of power consumption rates, however, the maximum range of these transporters is restricted to approximately 12,500 kilometers. Under special conditions, such as troop deployments, this range can be extended, but with significant impact on other power circuits. There are two emergency evacuation transporter complexes in the primary hull and two in the secondary hull.

4.16.02 MEDICAL

The Sickbay Complex has two transporters designed to accommodate patients and medical personnel. Larger, oblong pads can handle an anti-gravity gurney (or stretcher) and up to four attendants. Injured personnel can be delivered directly to the Sickbay from a planetary surface, another ship, starbase, or from any location within the ship itself. This capability can be crucial when serious injuries require immediate attention.

4.16.03 CARGO

Two large cargo transporters, both in the secondary hull, are used for inanimate objects. Each has the capability of handling up to five metric tons of material in a single transporter cycle. Use at this level, however, precludes the simultaneous operation of personnel transporters because of high power-consumption levels.

Two large cargo bays, adjacent to the cargo transporters, are designed to store non-replicable items. Other, smaller cargo storage facilities, are spread throughout the ship. Item are transferred from one hold to another by anti-gravity dollies.

5 SPECIFICATIONS

CLASSIFICATION	Starcruiser
CATEGORY	Cruiser
CLASS	GALILEO
TYPE	Class I
MODEL	Mark II
NUMBER PROPOSED	12
NUMBER CONSTRUCTED	12
NUMBER IN SERVICE	12

DIMENSIONS

Overall (in meters)
Length	358.65
Beam	164.38
Draft	95.27

Primary Hull (in meters)
Length	164.38
Width	164.38
Height	48.57

Secondary Hull (in meters)
Length	149.44
Width	66.37
Height	33.62

Warp Nacelles (in meters)
Length	177.51
Width	26.84
Height	17.66

DISPLACEMENT (in metric tons)
  Light   284,532.0
  Standard   305,000.0
  Full   340,528.0

PERFORMANCE

  Impulse (in seconds)
      $0.000—0.167c$ .330
      $0.168—0.334c$ .495
      $0.335—0.667c$ .660
      $0.668—0.999c$ .826

  Warp (in seconds)
      1.00—2.00   .244
      2.00—3.00   .391
      3.00—4.00   1.480
      4.00—5.00   2.128
      5.00—6.00   2.253
      6.00—7.00   2.434
      7.00—8.00   3.125
      8.00—9.00   4.468

SHIP'S COMPLEMENT
  Officers   304
  Enlisted   126
  Troops   68
  Passengers   184
  Emergency Conditions   +868

6 SYSTEM CONTRACTORS

Item	Designation	Contractor	Location
Builder		Albuquerque Division Sandia Ship-building and Conversion	Earth Orbit Spacedock Facilities, Earth (Sol)
Propulsion	SY-71/1-5AC Inter-phased Warp Engines	Leeding Engines, Ltd.	Sydney, Earth (Sol)
	"Power Six-Pack" Multi-ported Impulse Engines	Klorati Drives	Tellar (61 Cygni)
	"Eye of the Needle" Integrated Man-euvering Thrusters	Scarbak Propulsion Systems	Cairo, Earth (Sol)
Navigation	"Hawkeye III" Warp Celestial Guidance	Plessey Group	Essex, Earth (Sol)
Computers	Operations "Mega Monolithics" Tritonic V with Dayton Cross-multiplexer R-9 supplement	Monolithic Memories	Vulcan (40 Eridani)
	Scientific "Mega Monolithics" Tritonic V with Dayton Cross-multiplexer R-9 supplement	Monolithic Memories	Vulcan (40 Eridani)

Item	Designation	Contractor	Location
	Engineering "Montgomery II"	Monolithic Memories	Vulcan (40 Eridani)
	Fire Control "FasTrak"	Monolithic Memories	Vulcan (40 Eridani)
	Navigation "Mega Monolithics"	Monolithic Memories	Vulcan (40 Eridani)
	Main Library "Mega Monolithics"	Monolithic Memories	Vulcan (40 Eridani)
Sensors	OMNI-Synse Omni-directional, wide-spectrum, circum-ferentially mounted Sensor Package	Tachyon Industrial Division, Tachyon Micro-mechanics, Ltd.	Grindasa, Arcturus (Alpha Boötis)
Standard Phasers	RIM-18B (ECHO-enhanced) Independently Controlled Quad-mounts	Asakaze Ordnance Systems, Ltd.	Honshu-Hamamatsu, Earth (Sol)
Mega-Phasers	"King 60" (ECHO-enhanced) Multi-directional Cannon	Asakaze Ordnance Systems, Ltd.	Honshu-Hamamatsu, Earth (Sol)
Photon Torpedoes	MK-15, MOD5 Direct-connect Interphase Firing	Loraxial Magnetic Dynamics	Andor (Epsilon Indi)
Defense Systems	"Merlin II" Primary Force Field and Deflector Control System	Charlotte's Shields, Inc.	Quiberon Prime, Alpha Centauri (Al Rijil)

Item	Designation	Contractor	Location
	"Medusa" Weapon Fire Control System with Integrated Supplement	Bzevhistakis Kor Conglessum	Bortis, Tellar (61 Gygni)
	"Emperor" Integrated Cloaking Penetration and Stasis Countermeasures System	Shor Ta'kel Ltd.	Central Docks (40 Eridani)
Life Support	AE-5 Artificial Gravity Unit with Redundancy Module	Morris Gravitics Division, Morris Magtronics	Palynia, Mars (Sol)
	"Starcloak" Radiation Shielding	Tidjikal/Atar Associated Industries	Rastaribi, Regulus (Alpha Leonis)
	"Reuse II" Wast Regeneration System	Jullundur-Lahore, Ltd.	Bombay, Earth (Sol)
	"Golden Arch" Non-segregated, Multiple-supplemented Food Processing Units	McDonald's Division, Nutritech Corporation	Marsport, Mars (Sol)

System Contractors

SHIPBOARD ORGANIZATION

7 GENERAL INFORMATION

The operation of a large and complex entity such as a Starcruiser requires a hierarchial organization in order to function with efficiency and effectiveness. The Commanding Officer, as the senior individual aboard, is responsible for all facets of the ship's operation. Each Commanding Officer has a First Officer assigned who is immediately below the Commanding Officer and who is responsible for carrying out the desires and orders of the Commanding Officer.

The remainder of the crew is divided into seven departments, each composed of experts in a particular area. These departments (along with their associated abbreviation) are:

Command	CMD
Operations	OPS
Science	SCI
Engineering	ENG
Security	SEC
Logistics	LOG
Medical	MED

Within each department, with the exception of Medical, are divisions, composed of specialists in a particular field within that specific area.

Each department has a department head who is the senior officer in that particular field. Each division has a division officer, who controls the actions of assigned personnel. Refer to the organizational chart provided for clarity.

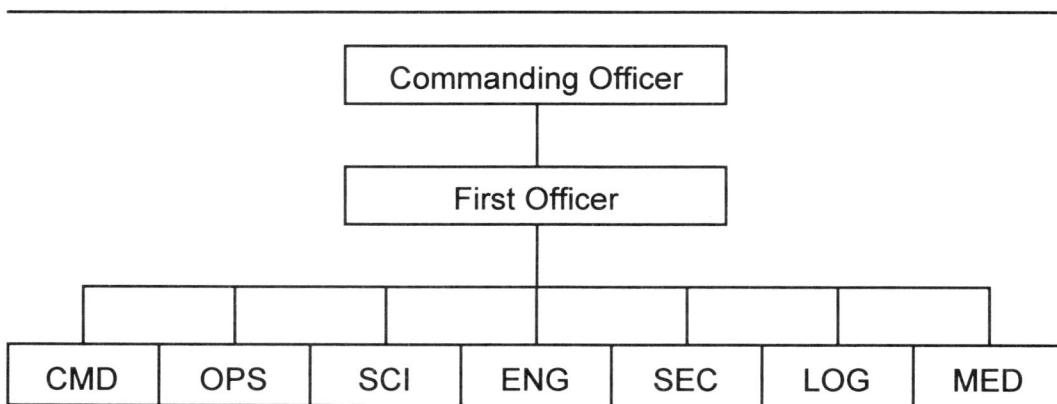

Starcruiser Organization Chart

Should an officer be unavailable or incapacitated, the next officer in the chain of command will assume those duties. The chain of command for GALILEO-class Starcruisers is:

> Commanding Officer
> First Officer
> Operations Officer
> Chief Science Officer
> Chief Engineering Officer
> Chief Security Officer
> Logistics Officer
> Chief Medical Officer

8 COMMAND DEPARTMENT

The Command Department is composed of individuals charged with the primary responsibility of overall guidance of the ship and its personnel; compliance with Starfleet regulations and directives, and with interpreting those regulations and directives as they apply to current and future ship operations and missions; shipboard administration and other special areas not appropriate to other departments or divisions aboard ship.

Although listed as a part of the Command Department, the Commanding Officer, as the senior officer aboard the Starcruiser, has no departmental obligations or responsibilities. The First Officer is the nominal head of the

Command Department. This department is divided into two divisions: Special Assistants and Administration. A discussion of their functions follow.

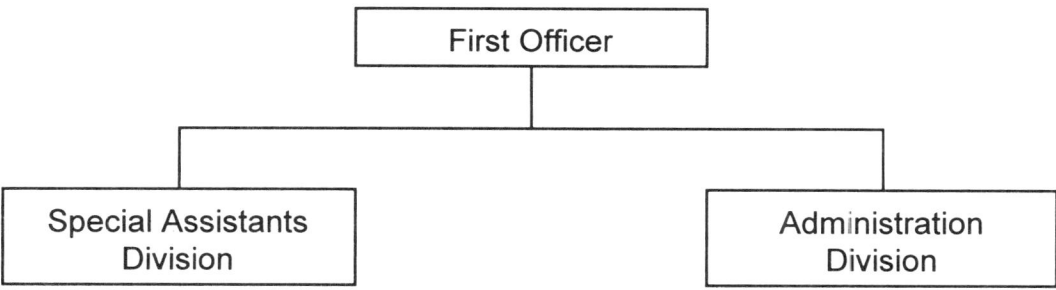

Command Department Organization

8.01 SPECIAL ASSISTANTS DIVISION

Because of the special missions for which the ship was designed, and the resultant deviations from normal Starfleet shipboard environments, a Ship's Orientation Officer has been assigned. The duties of this position are often accomplished in close cooperation with the personnel of the Medical Department.

The Religious Assistance Specialist has been trained to respond to the religious needs of the crew, and has been trained to recognize and cater to the requirements manifested by a large number of Federation religions. By special arrangements with the majority of the Federation's religious decision-making bodies, this person has been authorized to effect any of the ceremonies and rituals which may be required during an extended mission, such as marriages, divorces, baptisms, last rites, and other religion-specific duties. The Religious Assistance Specialist also works in close harmony with the Orientation Officer.

Two Diplomatic Services Specialists, both of whom are experienced diplomats, were specially trained by the Federation Diplomatic Corps to handle the unexpected situations which might arise in dealing with previously unknown species. After careful consideration, a decision was made by the Federation Diplomatic Corps and Starfleet Headquarters to absolve commanding officers of Starcruiser-class vessels of basic diplomatic responsibilities, although, as the senior officer present, the Commanding Officer has the final obligation to ensure diplomatic relations between the Federation and other non-aligned species are maintained in a professional and fair manner.

The First Contact Specialists assigned to the Special Assistants Division have the responsibility to apply the lessons learned from past First Contacts to any situation the Starcruiser-class vessel and its personnel may encounter.

In addition to these positions, a Ship's Historian/Archivist is assigned to maintain accurate and detailed verbal and non-verbal records of the ship's missions, situations, considerations, and results.

An Educational Services Officer is also assigned to assist personnel in furthering their individual knowledge, skills, and educational attainments during long voyages.

8.02 ADMINISTRATION DIVISION

Administrative specialists (called Yeomen) in this division maintain records, ensure accurate reports are made as scheduled, and perform other duties as assigned. Each department has at least one yeoman. Divisions which require dedicated administrative assistance (Shuttle Systems, Life Sciences, Physical Sciences, Social Sciences, and Space Sciences) are assigned a yeoman who is responsible specifically for that division.

The Captain's Yeoman, along with the Senior Yeoman, provide direct administrative support to the Commanding Officer and First Officer in consolidating and maintaining both internal and external administrative matters. They are also responsible for ensuring personnel records are up to date, scheduling and tracking personnel and promotion reviews, and departmental personnel assignments. Berthing assignments are handled by the Chief Administrative Officer.

The Senior Enlisted Advisor is tasked with counseling enlisted crew members in the areas of career, performance, and personal matters.

9 OPERATIONS DEPARTMENT

The Operations Department is charged with navigation and steering of the ship, communications within the ship and with those organizations such as Starfleet Headquarters which are located at a distance, with interpretation of unknown signals, and with operating and maintaining the numerous shuttle craft assigned to the ship for planetary use.

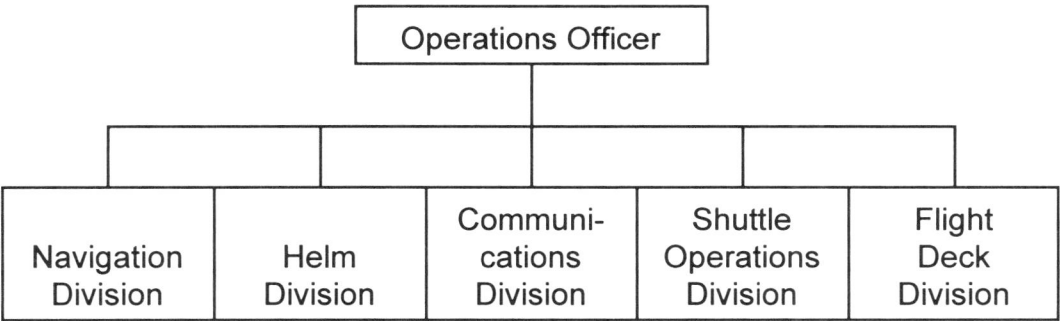

Operations Department Organization

9.01 NAVIGATION DIVISION

The Operations Computer aboard the ship is the primary source for position updates, chronometer calibration, and course determinations. It uses navigation and time beacons, sensor inputs, and complicated algorithms to ensure that the navigator is presented with precise, up-to-date data and course plots. It is imperative, however, that these data and course projections be independently verified.

Four navigators are assigned to this division, and each stands watches on the Main Bridge. They are responsible to the Captain, First Officer, or Officer-of-the-Deck (as appropriate) during their watch. Their duties are to:

(1) calculate and update the ship's present position with regard to velocity and maneuvers;

(2) locate the coordinates of the ship's designated destination using the ship's operations computer;

(3) calculate the best course from the origin coordinates to the destination coordinates, allowing for intervening Neutral Zones and catalogued hazards to navigation;

(4) act as the auxiliary Weapon's Officer whenever the Bridge Tactical Console is unmanned using the Navigation Console Fire-Control Panel.

The three Navigational Systems Analysts assigned to the division update navigational charts used by the operations computer and the Navigation Console on the Main Bridge and the Auxiliary Bridge based upon data acquired by the lateral sensor array. Update data provided by other ships via starbases (provided on an as-occurring basis) is entered into the navigational system database when appropriate.

Maintenance, repair, and upkeep on navigational system components and equipment falls under the purview of the three Navigational Systems Technicians. One technician is assigned to each shift to ensure continual coverage around the clock.

9.02 HELM DIVISION

Although the majority of the systems aboard the ship are automated, it is important that an experienced helmsman be available to respond to unforeseen situations and to provide that spontaneity and innovation available only from a manned helm.

Four helmsmen are assigned to this division, and each one stands watches on the Main Bridge. They are responsible to the Captain, First Officer, or Officer-of-the-Deck (as appropriate) during their watch. Their duties are to:

(1) execute any course stored in the Preset Course Memory Bank at the speed designated by the Commanding Officer, First Officer, or Officer-of-the-Deck (as appropriate);

(2) monitor the ship's progress along its course, alerting the Commanding Officer, First Officer, or Officer-of-the-Deck (as appropriate) to any uncharted

objects or hazards, and to override the existing course as directed so as to avoid the hazard or stop for investigation; and,

(3) maneuver the vessel on manual control for such purposes as evasive maneuvers, intercept courses, orbital insertions, or other maneuvers requiring precise, non-automated control.

Helm Systems Analysts and Helm Systems Technicians maintain and repair the helm console and its associated equipment. One Analyst and one Technician is assigned to each shift.

9.03 COMMUNICATIONS DIVISION

The ability to transmit and receive messages from other ships (Federation or otherwise), Starbases and planets is an important facet of shipboard operations. Internal communications through the intercom network is also of paramount importance. Ensuring this ability to communicate is the task assigned to the Communications Division.

The ship's external communications equipment is handled by Subspace Equipment Specialists, who report to the Communications Systems Officer. They are also responsible for updating the Universal Translator as necessary.

Internal Equipment Specialists perform repair and maintenance actions on the ship's intercom system and maintain all tricorders aboard the ship. Equipment Technicians assist in this task as required.

9.04 SHUTTLE OPERATIONS DIVISION

There are eight shuttlecraft attached to the ship: four standard shuttles, one medical shuttle, one aquatic shuttle, and two WorkBee craft.

Each shuttlecraft is assigned a dedicated pilot. WorkBee craft are piloted by either a shuttlecraft pilot or another person on the ship with specific qualifications.

Assignment of shuttlecraft to specific missions is the responsibility of the Commander, Shuttlecraft Fleet, with the caveat that a specific shuttlecraft and pilot may be requested by either the Commanding Officer or the Officer-in-Charge of a particular planet survey or expedition.

Mechanics and Support Technicians repair and maintain the shuttlecraft as necessary. Each is provided instruction in Repair and Maintenance Bay operations which, because of variable gravity considerations, can be hazardous.

9.05 FLIGHT DECK DIVISION

Although the operation and maintenance of the ship's shuttles are assigned to personnel of the Shuttle Operations Division, the safe and efficient operation of the Hangar Deck and all objects therein falls to the personnel of the Flight Deck Division.

Movement of any craft within the Hangar Bay (or Flight Deck) is controlled by the Chief Flight Deck Officer and the Flight Deck Control Specialists. This includes craft from another ship or starbase landing or launching, and movement from and to the Repair and Maintenance Bay.

The Hangar Bay doors delineate the point where control is transferred to either the Flight Deck Officer or to the Pilot-in-Command. As soon as an incoming craft crosses the line of the Hangar Bay doors, control is passed to the Flight Deck Officer. As soon as an outgoing craft crosses the line of the Hangar Bay doors, control is passed to the Pilot-in-Command.

10 ENGINEERING DEPARTMENT

Responsibilities of the Engineering Department include both the proper operation and maintenance of impulse and warp drive propulsion, life-support systems, transporter systems, ship defensive/offensive weapons systems (phasers, Mega-phasers, and photon torpedo capabilities), the deflector system, the cloaking penetration system, the tractor beam units, and auxiliary systems throughout the ship.

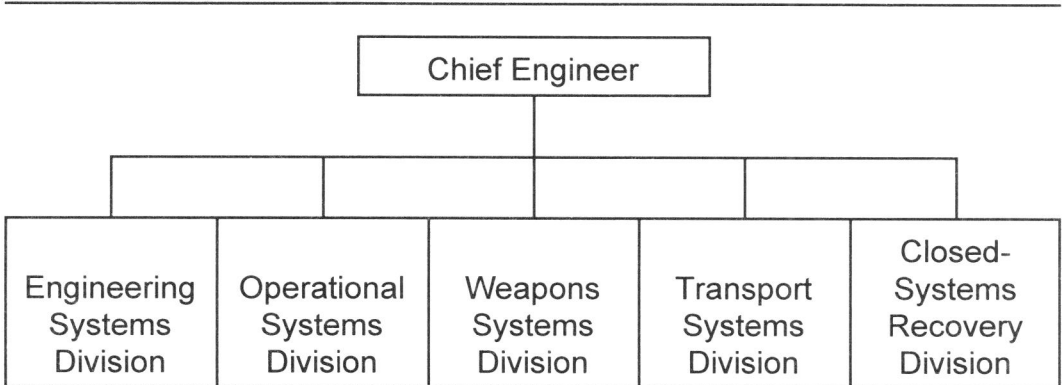

Engineering Department Organization

10.01 ENGINEERING SYSTEMS DIVISION

Engineers assigned to this division have the responsibility for maintaining, repairing and calibrating the warp drive, the impulse drive system and the ship's thrusters.

Matter/Anti-matter Specialists work closely with personnel of the Operational Systems Division to ensure the effective and efficient operation of the ship's power sources. Dilithium Crystal experts monitor the status and consumption rates of this critical component of the propulsion system. Maintenance Technicians provide repair and upkeep assistance to the Engineers and Specialists in this Division.

10.02 OPERATIONAL SYSTEMS DIVISION

With the exception of weapons and transporters, all other engineering systems on the ship fall under the purview of the Operational Systems Division. These systems include life-support, environmental, deflector shields, the cloaking penetration system, tractor beams, sensor packages, and auxiliary systems.

The Inertial Dampening System is monitored by the Structural Engineering Specialist. The Acoustical Systems Technician assists personnel in the Communications Division to maintain the ship's internal communications systems. Life-Support System Technicians, Environmental Systems

Technicians, and the Bio-Systems Specialist are involved with artificial gravity generators, atmospheric quality regulators, and humidity and temperature controllers throughout the ship.

Deflector Shield Specialists maintain and repair the main force field generators which provide protection to the ship and the Navigational Deflector system. Tractor Beam Specialists work on the repulsor/attractor units located at various points on the exterior hull.

Power generation, power supply conduits and power loading/sharing circuits are controlled by Auxiliary Systems Specialists. The OMNI-Synse lateral sensor array and the sensor packages on probes and torpedoes are handled by Sensor Systems Specialists.

10.03 WEAPONS SYSTEMS DIVISION

GALILEO-class Starcruisers are the most powerful weapon platforms in the Fleet. Maintenance, calibration, and repair of the ship's phaser system, Mega-phaser cannons, photon torpedoes and torpedo launchers, and the Probe Launch Room is accomplished by the Specialists and Technicians in this division.

10.04 TRANSPORT SYSTEMS DIVISION

There are several transporters located throughout the ship including standard six-person units, emergency 23-person transporters, cargo transporter units located near the cargo bay in the secondary hull, and medical transporters located in or near the Medical Complex. These transporter complexes are operated, maintained, and repaired by this division.

10.05 CLOSED-SYSTEMS RECOVERY DIVISION

Like most large, deep-space vehicles, the GALILEO-class Starcruiser sustains a closed ecological system to maintain environmental support.

Unlike a planetary biosphere, however, a starship must use technologic means to approximate the complex ecologic processes that sustain life.

Among these processes are the waste management and recovery systems, which make optimal reuse of waste products possible. The personnel assigned to this division ensure these waste management and recovery systems are maintained at a high level of operational efficiency in order to make maximum utilization of on-board ecological materials.

Refurbishers repair and maintain decks, bulkheads, and overheads throughout the ship. They also make desired modifications to compartments and staterooms. Janitorial Services personnel inspect the ship on a regular basis to ensure cleanliness of compartments and spaces, assisting as necessary to meet the high standards demanded on a star ship.

Systems to treat and recycle waste products and all heads on the ship are the responsibilities of Sanitation Maintenance Technicians. The Botanical Garden and the organic plant-growth complexes are monitored and maintained by Hydroponics Specialists.

There are two main fabrication systems on board the ship: the food synthesizers and the hardware fabricators. This division is responsible for maintaining and operating the hardware fabricators, which operate at a lower resolution level than the food synthesizers, and are used to replicate those items which are used on a daily or as-needed basis in such areas as medical, engineering, and crew's quarters.

11 SCIENCE DEPARTMENT

The largest department aboard the GALILEO-class Starcruiser, the Science Department contains those specialists around whom the ship was designed. A large percentage of the space inside the ship is dedicated to laboratories and research facilities where these scientists work.

The department is divided into four divisions, each of which is charged with the responsibilities for a different aspect of scientific study.

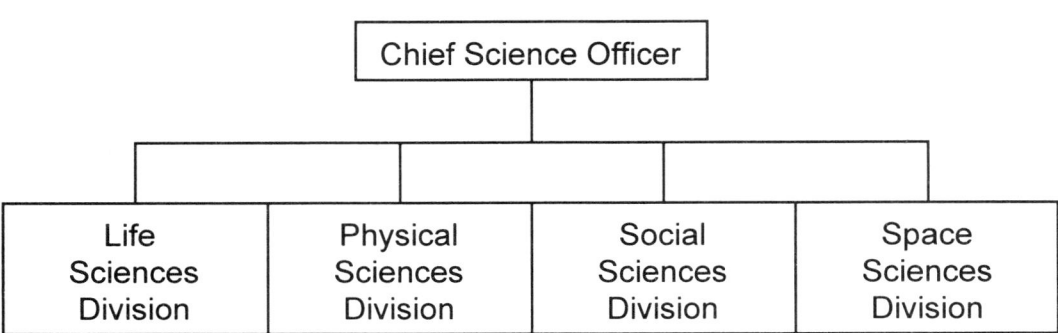

Science Department Organization

11.01 LIFE SCIENCES DIVISION

The discovery of life in another environment, whether it be on a planet or in space, is an exciting event, regardless of its complexity. Scientists assigned to this division are called into play for the search for and study of new and existing life forms.

The scientists assigned to this division and their areas of expertise are as follows:

Agronomist: study the mass development of plants by deriving new growth methods or by controlling diseases, pests and weeds

Anatomist: study the structure of organisms from cell structure for formation of tissues and organs

Biochemist: research the chemistry of life processes in plants and animals

Bioengineer: investigate the application of engineering science and technology to problems of biology and medicine

Biologist: investigate the origins, history, physical characteristics, life process and habits of land plants and animals

Biotechnologist: use of the data and techniques of engineering and technology for the study and solution of problems concerning living organisms

Botanist: study plants, their life, structure, growth, classification, etc.

Cryogenics Officer: examine the production of very low temperatures and their effects on organic and inorganic substances

Ecologist: explore the relationship between organisms and their environments

Embryologist: investigate the development of an animal from a fertilized egg through the eventual birth

Entomologist: study of insects (and like organisms) and their relation to plant and animal life

Geneticist: deal with heredity and variation in similar or related animals and plants

Herpetologist: study reptiles and amphibians

Horticulturalist: works with orchard and garden type plants

Ichthyologist: analyze aquatic animals, their structure, classification, and life history

Limnologist: study fresh water aquatic life

Mastologist: study of warm-blooded, usually hairy, vertebrates whose offspring are fed with milk secreted by the female mammary glands

Marine Biologist: investigate the origins, history, physical characteristics, life process and habits of aquatic plants and animals

Microbiologist: deal with microorganisms such as bacteria, viruses and molds

Ornithologist: study the origins, history, physical characteristics, life process and habits of air-capable animals

Paleontologist: investigate pre-historic life forms through the study of fossils

Pathologist: examine the effects of diseases, parasites and insects on cells, tissues, and organs

Toxicologist: study the effects of drugs, gases, poisons, and other substances on the functioning of tissues and organs

Zoologist: study origin, behavior, and life processes of animals

11.02 PHYSICAL SCIENCES DIVISION

A wide range of specialists who concentrate on investigating planetary surfaces are assigned to this division. Land surfaces, including minerals, metals, landforms, and geological formations, are studied by a number of geological and geophysical scientists. A planet's liquid areas come under the scrutiny of oceanographers and hydrologists. Also contributing to the research capabilities of the Science Department are chemists, mathematicians and computer systems analysts.

The scientists assigned to this division and their areas of expertise are as follows:

Analytical Chemist: determine the structure, composition, and nature of substances and develop new techniques

Cartographer: prepare charts and maps of planetary bodies

Chemical Oceanographer: study the chemical composition of ocean water and sediments

Climatologist: investigate general trends of climate on planetary bodies

Economic Geologist: locate various minerals of societal value, such as dilithium, etc.

Exploration Physicist: use various seismic and sensoric prospecting techniques to locate valuable minerals

Geochemist: study the chemical composition and changes in minerals and rocks

Geochronologist: determine age of geological formations by the radioactive decay of their elements

Geodesist: explore the size, shape, and gravitational fields of planets

Geological Oceanographer: study the ocean's underwater mountain ranges, rocks, and sediments

Geomagnetician: investigate planetary magnetic fields

Geomorphologist: study forces such as erosion and glaciation

Hydrologist: investigate the distribution, circulation, and physical properties of underground and surface waters

Inorganic Chemist: study compounds other than carbon

Mineralogist: analyze and classify minerals and gems by composition and structure

Organic Chemist: study the structure of all carbon compounds

Paleomagnetician: determine past magnetic fields from rocks or lava flows

Paleontologist: study plant and animal fossils found in geological formations

Palynologist: locate liquid deposits by studying tiny organic fossils

Physical Meteorologist: research the physical characteristics of atmospheres

Physical Oceanographer: research the physical properties of the ocean such as waves, tides, and currents

Planetologist: investigate the composition and atmospheres of moons, planets, and other bodies in a star system

Seismologist: study planet's interior and vibrations caused by natural explosions

Soil Scientist: analyze characteristics of life-bearing soil

Stratigrapher: study the distribution and arrangement of sedimentary rock layers

Synoptic Meteorologist: study planetary weather as differentiated from climate

Volcanologist: study active and inactive volcanoes

11.03 SOCIAL SCIENCES DIVISION

Scientists in this division study the interactions of social groups. Various kinds of experts within this division can study the past relations of both viable and extinct civilizations. A myriad of information can be revealed by looking at a society's geographical, medical, art, and infrastructural accomplishments. If an extinct civilization is discovered, these personnel become the leaders of the investigation.

The scientists assigned to this division and their areas of expertise are as follows:

Anthropologist: study the variety, physical and cultural characteristics, distribution, customs, social relationships, etc., of life forms

Archeologist: study the life and culture of past life forms as by excavation of ancient cities, relics, artifacts, etc.

Curator: catalogs, classifies, stores, and preserves artifacts

Demographic Researcher: study the distribution and interaction of life forms in their native environments

Ethnoscientist: deal descriptively with cultures, especially currently viable societies

Geographer: deal descriptively a planet's surface, its division into continents, and the climate, plants, animals, natural resources, inhabitants, and industries of the various divisions

Historian: develop systematic accounts of what happened in the life or development of a life form in chronological order with analysis

Legal Specialist: deal with the development, practice, and interrelationships of laws between and among life forms

Paleoanthropologist: study the variety, physical and cultural characteristics, distribution, customs, social relationships, etc., of ancient life forms

Political Scientist: correlation of political institutions and the principles, organization, and methods of government

Sociologist: study the organization, needs, and development of societies

11.04 SPACE SCIENCES DIVISION

From quasars to super-novas, from multiple-star systems to black holes, these experts spend long hours at sub-light speeds collecting data on a variety of phenomena. During faster-than-light travel, they are able to concentrate on interpreting the data gathered by the sensor package. While in orbit around a planet, specialists in this division concentrate on studying weather patterns and their effects.

The scientists assigned to this division and their areas of expertise are as follows:

Aeronautical Specialist: explore the design, development, and operation of craft capable of flight only within a planet's atmosphere

Aerospace Specialist: explore the design, development, and operation of craft capable of flight in or outside a planet's atmosphere

Astronautical Specialist: explore the design, development, and operation of craft capable of flight only outside a planet's atmosphere

Astronomer: study stars, planets, etc., including their origins, evolution, composition, motions, relative positions, sizes, etc.

Astronomical Cartographer: prepare and update holographic charts used for navigation; works closely with personnel in the Navigation Division of the Operations Department

Astrophysicist: deal with the physical properties of the universe, including luminosity, density, temperature, and chemical composition

Cosmologist: study the form, content, organization, and evolution of the universe

Elementary Particle Physicist: deal with the properties, changes, interactions, etc., of matter and energy on the most basic level

Molecular Physicist: deal with the properties, changes, interactions, etc., of matter and energy on the molecular level

Nuclear Physicist: deal with the properties, changes, interactions, etc., of matter and energy on the atomic level

Plasma Physicist: deal with the properties, changes, interactions, etc., of high-temperature, ionized gases composed of electrons and positive ions

Pulsar Specialist: study of rotating neutron stars which emit electromagnetic radiation at short and very regular intervals and used for navigation

Quasar Specialist: study of starlike bodies which emit immense quantities of light and radio waves

Thermodynamic Physicist: investigate the transformation of heat to and from other forms of energy

12 MEDICAL DEPARTMENT

The GALILEO-class Starcruiser has, as its primary mission, the investigation of areas which may or may not lie within recognized Federation space. As a result, the possibilities for crew injuries is directly proportional to the distance from planet or starbase medical facilities. Therefore, the medical expertise aboard the ship is capable of handling any medical situation.

The Medical Department is the only department on the ship that is not broken down into divisions. As a small group of people with extremely integrated duties, it was determined that a single entity would work best.

Invasive techniques, although no longer a mainstay of Federation medicine, are still an important aspect of maintaining the health of crew members or responding to massive traumas. Physicians are assigned in order to be prepared for any contingency. Nurses, laboratory technicians and maintenance personnel ensure that the mission of this department is carried out in the most efficient and effective manner possible.

During Red Alerts, Medical Department personnel are assigned to Main Sickbay, Secondary Sickbay, Mobile Emergency Medical Teams (MEMT), and the Main Lounge. For a listing of Red Alert assignments, see the SMD.

Mobile Emergency Medical Teams, composed of a physician, a nurse, and other medically trained personnel, are stationed at various locations throughout the ship. Should injuries occur during a Red Alert, the closest MEMT will respond and provide appropriate care to the injured who do not require transport to Main or Secondary Sickbay.

The Main Lounge is converted into a Triage Center, where casualties are categorized as to their level of required medical care. Minor injuries are treated, while more serious cases are sent to Secondary Sickbay or Main Sickbay, depending upon the severity.

13 SECURITY DEPARTMENT

Internal security of the ship falls within the purview of the Security Department. All ships have areas of restricted access or occasions when individual liberty may have to be curtailed. The personnel of this department have the responsibility of ensuring that these duties are performed in a expeditious manner.

In addition, Security Department personnel maintain the ship's personal weapons and also train shipboard personnel in their appropriate use.

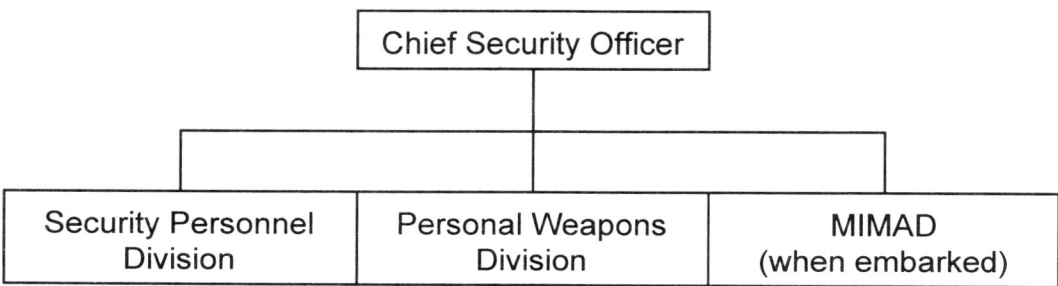

Security Department Organization

13.01 SECURITY PERSONNEL DIVISION

Certain areas of the ship are inherently dangerous to unfamiliar personnel. In order to prevent accidents which may injure individuals or perhaps damage critical equipment, the Security Personnel Division is charged with protecting those areas from unauthorized entry. In addition, there may be cases in which individuals must be isolated from the remainder of the crew—this also is the responsibility of this division.

The Security Personnel Division operates independently of the ship's routine. Security Officers are assigned to one of three Security Teams, each headed by a Security Watch Officer. During normal shipboard situations, roving patrols move constantly throughout the ship. During other Security Conditions, spaces are guarded by Security personnel to prevent un-authorized access. For a more detailed description of how this Division operates, see Page 146.

13.02 PERSONAL WEAPONS DIVISION

All hand weapons aboard the ship are maintained, repaired, and inventoried by personnel in this division. Training in the use of weapons is conducted for the crew, as is hand-to-hand combat techniques.

14 MOBILE INDEPENDENT MARINE ACTION DETACHMENT

In carrying out the offensive or defense mission of the ship, a Starcruiser is considered to be a Fleet-Action Coordination vessel. This designation carries with it a secondary mission of transporting assault troops. To fulfill this role, provisions have been made to berth a Marine Detachment.

When required, the ship will take on a contingent of Marines from one of the Independent Marine Action Groups (or IMAG). Including the Major in charge of the Detachment and the Detachment Assistant (a Sergeant Major), 68 Marines make up a Mobile Independent Marine Action Detachment (MIMAD).

The Detachment is divided into three platoons: Light Weapons, Heavy Weapons, and Reconnaissance. Each platoon is further separated into three sections of three teams per section.

The Light Weapons Platoon is utilized when a highly mobile, fast-reaction unit is required to assault a target which is not heavily defended. This unit is also used in hostage or perimeter defense scenarios.

Phaser rifles, grenade launchers, and other high-energy weapons comprise the arsenal carried by the Heavy Weapons Platoon. They are used when a target is heavily defended or when a shock attack is necessary.

The Reconnaissance Platoon is trained in covert operations and intelligence gathering. They carry only light weapons (usually only hand phasers) and sophisticated but light-weight communication-interdiction equipment. These troops are equipped and trained for night operations.

Special training in perimeter defense, rapid deployment, and transporter dispersal techniques make the MIMAD aboard a Starcruiser the most intensely trained and highly capable troops in Starfleet Marines.

15 LOGISTICS DEPARTMENT

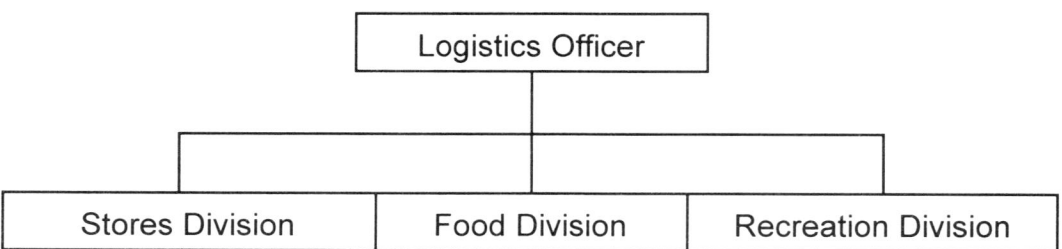

Logistics Department Organization

15.01 STORES DIVISION

While fabrication of spare parts, food, and clothing is a common occurrence aboard the ship, there are cases in which items are used in such a sufficient quantity that it is more economical to store finished products than to spend the energy to carry raw materials and synthesize the finished product on demand.

Additionally, significant stores of critical spares and consumables are maintained for possible use during Red Alert situations when power for replication or fabrication systems may be severely restricted or unavailable. Stores Division personnel store, maintain, inventory, and deliver such supplies when and where necessary.

15.02 Food Division

Replication of food aboard the ship is standard, utilizing the non-segregated, multiple-supplemented food processing units developed by McDonald's Division of Nutritech Corporation located throughout the ship. However, each unit requires maintenance and, occasionally, repair.

In addition to these responsibilities, the personnel of this division ensure that specialty foods required by non-humanoid crew members, VIPs, or patients in Sickbay, are prepared correctly and available as needed.

15.03 Recreation Division

The extended cruises upon which the ship may be sent at any time can result in a lowering of crew efficiency from a lack of activities. The Recreation Division is composed of individuals from a number of varied and diverse backgrounds charged with providing the crew with a number of diversions.

Actors, painters, sculptors, physical activities coaches and music experts are available for the crew to either improve or develop expertise in their leisure time. In addition, the several lounges located throughout the ship are manned by Lounge Services Personnel in order to provide "the personal touch."

SHIPBOARD OPERATIONS

16 Daily Routines

GALILEO-class Starcruisers, like all other starships in Starfleet, operate on an around-the-clock basis. Two types of activities occur: scheduled and unscheduled. Both are listed by commencement times and, if applicable, termination times, in a Plan of the Day, which can be accessed by any individual or station console. Items included in the Plan of the Day are updated at the end of each watch so as to provide accurate information to on-coming watch personnel and off-duty personnel.

Scheduled activities are those activities which occur at the same time each day. These include: sounding ship's bells, changing the watch, commencing required Level 1 maintenance checks, setting the Day Watch and the Night Watch, and making a Daily Report to the Officer-of-the-Deck (reporting the status of all equipment, spaces, and personnel in the respective department).

Unscheduled activities are those not scheduled on a recurring basis. Some items which fall into this category include: equipment, personnel, or space inspections; special celebrations such as wedding or birthdays; award or promotion ceremonies; docking or undocking operations; arrival or departure at planets, starbases, or other specified locations; and, shuttle flight operations. Other unscheduled activities included in the Plan of the Day are VIP arrivals and/or departures, individual- or team-sport competitions, and musical and theatrical performances.

Two other items are published in the Plan of the Day: the Binnacle List, a status of any injured or ill personnel; and an Out-of-Commission Advisory, a listing of any equipment or spaces which cannot be utilized due to malfunction or maintenance.

A listing of all scheduled activities by time is provided below.

Time	Activity
0030	Make 1 bell over appropriate ship circuits.
0100	Make 2 bells over appropriate ship circuits.
0130	Make 3 bells over appropriate ship circuits.
0200	Make 4 bells over appropriate ship circuits.
0230	Make 5 bells over appropriate ship circuits.
0300	Make 6 bells over appropriate ship circuits.
0330	Make 7 bells over appropriate ship circuits.
0400	Make 8 bells over appropriate ship circuits.
0430	Make 1 bell over appropriate ship circuits.
0500	Make 2 bells over appropriate ship circuits.
0530	Make 3 bells over appropriate ship circuits.
0600	Make 4 bells over appropriate ship circuits.
0630	Make 5 bells over appropriate ship circuits.
0700	Make 6 bells over appropriate ship circuits.
0730	Make 7 bells over appropriate ship circuits. Update Plan of the Day in Main Library Computer.
0800	Make 8 bells over appropriate ship circuits. Relieve the watch. Set Day Watch. Conduct required maintenance.
0830	Make 1 bell over appropriate ship circuits.
0900	Make 2 bells over appropriate ship circuits.
0930	Make 3 bells over appropriate ship circuits.
1000	Make 4 bells over appropriate ship circuits.

Time	Activity
1030	Make 5 bells over appropriate ship circuits.
1100	Make 6 bells over appropriate ship circuits.
1130	Make 7 bells over appropriate ship circuits.
1200	Make 8 bells over appropriate ship circuits.
1230	Make 1 bell over appropriate ship circuits.
1300	Make 2 bells over appropriate ship circuits.
1330	Make 3 bells over appropriate ship circuits.
1400	Make 4 bells over appropriate ship circuits.
1430	Make 5 bells over appropriate ship circuits.
1500	Make 6 bells over appropriate ship circuits.
1530	Make 7 bells over appropriate ship circuits. Update Plan of the Day in Main Library Computer.
1600	Make 8 bells over appropriate ship circuits. Relieve the watch. Conduct required maintenance.
1630	Make 1 bell over appropriate ship circuits.
1700	Make 2 bells over appropriate ship circuits.
1730	Make 3 bells over appropriate ship circuits.
1800	Make 4 bells over appropriate ship circuits.
1830	Make 5 bells over appropriate ship circuits.
1900	Make 6 bells over appropriate ship circuits.
1930	Make 7 bells over appropriate ship circuits.

Time	Activity
2000	Make 8 bells over appropriate ship circuits. Set the Night Watch. Department Heads made daily reports to the Officer-of-the-Deck on the Main Bridge.
2030	Make 1 bell over appropriate ship circuits.
2100	Make 2 bells over appropriate ship circuits.
2130	Make 3 bells over appropriate ship circuits.
2200	Make 4 bells over appropriate ship circuits.
2230	Make 5 bells over appropriate ship circuits.
2300	Make 6 bells over appropriate ship circuits.
2330	Make 7 bells over appropriate ship circuits. Update Plan of the Day in Main Library Computer.
2400	Make 8 bells over appropriate ship circuits. Relieve the watch. Conduct required maintenance.

16.01 TIMEKEEPING

The natural habitat of a star ship is deep space, where there is no planetary rotation to provide a convenient method of measuring the passage of time. This lack of a constant and reliable method of indicating the time has proved to be a major difficulty aboard deep space vessels for all known sentient beings. In order to provide such a measure of consistency to the crew of the Starcruiser, an artificial 24-hour day has been created. Tied to the standard kept at the London Observatory, London, England, Earth, Universal Standard Time has been adopted as the criteria for measuring time on all GALILEO-class Starcruisers.

The 24-hour clock is also used to eliminate any possible confusion in establishing the time of an occurrence, whether current or scheduled. The hours begin at 0030 (12:30am) and count up through 2400 (midnight).

To mark the passage of time aboard the ship, a system of "bells," first established on 18th century Earth sailing ships, is used. Beginning at 0030 with one tone (which sounds similar to the ring of a bell), each half hour of the day and night is marked by an increasing number of tones. The number of "bells" sounded, after eight bells, begins again at one. Therefore, each half hour is indicated by an odd number of bells, and each hour is marked by an even number of bells. All circuits listed on Page 156 will carry these bells. A convenient chart to clarify the number of bells and their equivalent times is provided on Page 143.

16.02 WATCHES AND SHIFTS

With the exception of the Commanding Officer, First Officer, and Department Heads, every officer and enlisted person aboard the ship is assigned to a shift section. There are three shift sections, each having an equal portion of personnel from each Department or Division.

Section	Weeks 1-6	Weeks 7-12	Weeks 13-18
ALPHA	2400—0800	0800—1600	1600—2400
BETA	0800—1600	1600—2400	2400—0800
GAMMA	1600—2400	2400—0800	0800—1600

Shift Rotation Schedule

Each of these sections (Alpha, Beta, Gamma) takes a turn at each of the three watches, rotating on a six-week cycle as shown below. At any time First Shift is that section on duty. Second Shift is that section which will relieve First Shift, and Third Shift is that section which was relieved by First and which will eventually relieve Second Shift. Each shift is eight hours long.

24-hr Clock	Equivalent	Bells	24-hr Clock	Equivalent	Bells
0030	0:30AM	1	1230	12:30PM	1
0100	1:00AM	2	1300	1:00PM	2
0130	1:30AM	3	1330	1:30PM	3
0200	2:00AM	4	1400	2:00PM	4
0230	2:30AM	5	1430	2:30PM	5
0300	3:00AM	6	1500	3:00PM	6
0330	3:30AM	7	1530	3:30PM	7
0400	4:00AM	8	1600	4:00PM	8
0430	4:30AM	1	1630	4:30PM	1
0500	5:00AM	2	1700	5:00PM	2
0530	5:30AM	3	1730	5:30PM	3
0600	6:00AM	4	1800	6:00PM	4
0630	6:30AM	5	1830	6:30PM	5
0700	7:00AM	6	1900	7:00PM	6
0730	7:30AM	7	1930	7:30PM	7
0800	8:00AM	8	2000	8:00PM	8
0830	8:30AM	1	2030	8:30PM	1
0900	9:00AM	2	2100	9:00PM	2
0930	9:30AM	3	2130	9:30PM	3
1000	10:00AM	4	2200	10:00PM	4
1030	10:30AM	5	2230	10:30PM	5
1100	11:00AM	6	2300	11:00PM	6
1130	11:30AM	7	2330	11:30PM	7
1200	NOON	8	2400	MIDNIGHT	8

The six-week rotating cycle was arrived at after extensive studies and analyses of optimum biorhythm patterns among humanoids. Should this rotating cycle adversely affect a non-humanoid race, alternative shift assignments can be made with the concurrence of the Chief Medical Officer and the individual's respective Department Head.

Department Heads are not assigned to shift sections. Although they must serve one watch in three, they are permitted to adapt their scheduling so as to be present at their duty stations for critical events. The Chief Medical Officer pays special attention to these officers so that they do not become exhausted through biorhythm upset.

A frequent exception to the concept of one-third of a Department's personnel being on-duty during each of the three watches is the Science Department. As some work requires the combined efforts on behalf of all Science personnel, it is normal for all Science personnel to work during the same watch, usually 0800—1600.

On the other hand, some experiments or research might require around-the-clock supervision, or be required before a certain deadline. In such cases, the Chief Science Officer will coordinate and control such revisions to the duty roster as may be necessary.

17 ALERT CONDITIONS

The following Alert Status Standing Orders refer to all personnel except the Security Department. The Alert Status controls the readiness and deployment of the ship's complement for battle or other emergency conditions.

First Shift refers to the shift presently on duty, Second Shift refers to the next shift scheduled to relieve the First, and Third Shift refers to the shift just relieved by the First.

17.01 SHIPBOARD

Based upon the actual or expected situation which may exist aboard ship, certain actions are required by personnel. The following paragraphs define the various Alert conditions and the responses necessary to maintain stations at a level to react appropriately.

17.01.01 GREEN

This is the normal cruising status of the ship. No dangerous conditions exist on board or in the immediate vicinity of the ship. This condition may be instituted by the Commanding Officer, First Officer, or Officer-of-the-Deck.

First Shift on duty, Second and Third Shifts off duty.

17.01.02 YELLOW

A hazardous or potentially hazardous condition exists either on the ship or in the immediate vicinity. This condition may be implemented by the Commanding Officer, First Officer, or Officer-of-the-Deck.

First Shift remains on duty and prepares the Duty Station for possible danger (warming-up equipment that will be needed, storing and securing fragile items, etc.). Second Shift goes on duty and reports to respective Duty Stations, joining First Shift. Each station reports to the Bridge when the Duty Station is manned and ready.

17.01.03 RED

Known also as Battle Stations or General Quarters. Either a life-threatening condition exists on the ship or in the immediate vicinity, or the ship is in an offensive or defense situation which may result in the use of weapons and possible damage to the ship. May be implemented by the Commanding Officer or the First Officer.

Third Shift goes on duty and reports to their respective Duty Stations joining First and Second. Each Duty Station reports to the Bridge when the station is fully manned. Personnel assigned to Damage Control Parties will prepare for personnel injuries and possible catastrophic system failures.

17.01.04 Abandon Ship

No possibility exists that the vessel can continue to provide life support for the crew because of critical systems failures. All personnel proceed immediately to lifeboats or other assigned evacuation stations. Auxiliary craft will be launched immediately.

17.02 Security

Because of the unique responsibilities of the Security Department personnel, they are subject to a different set of regulations from the remainder of the crew. Personnel of the MIMAD are considered part of the Security Department for purposes of this paragraph.

First Shift is that section presently on duty. Second Shift is that section which will relieve First Shift and be relieved by Third Shift. Third Shift is that section which will relieve Second Shift and be relieved by First Shift.

There are four Security Conditions. The actions to be taken and the authorization for each are listed below.

17.02.01 Condition 4

This is the normal steaming condition for Security Department. First Shift is on duty, Second and Third Shifts are off duty. All Security personnel are armed with phasers set on STUN.

Two sentries are assigned to control access to the Auxiliary Bridge and to the Brig (if occupied). Three roving patrols of two Security Officers each ensure the safety of all personnel. One patrol makes random sweeps of the primary hull, another patrol makes random inspections of all spaces in the secondary hull, and the third patrol makes random checks throughout the entire ship.

When the MIMAD is embarked, additional areas where Marine sentries are posted include the Main Bridge, all Engineering spaces, all Transporter complexes, the Shuttlebay, designated Cargo Holds, air locks, and all weapons compartments.

17.02.02 CONDITION 3

Security Condition 3 is set for the Security Department personnel any time dignitaries are present on the ship and can be implemented by the Chief Security Officer, Officer-of-the-Deck, First Officer or Commanding Officer.

The definition of dignitaries includes planetary representatives (such as Ambassadors), whether members of the United Federation of Planets or not, and Flag Officers.

Two sentries will be posted outside assigned quarters and two Security Officers will be present at all functions, meetings, or negotiations attended by these dignitaries.

Additional personnel required to fulfill these responsibilities will be drawn from other shifts not on duty for as long as the requirements exists.

When embarked, the MIMAD will be deployed as directed by the Detachment Commanding Officer.

Security personnel will maintain their phasers on HEAVY STUN.   Deadly force is NOT authorized under Condition 3 unless specifically directed by the Commanding Officer.

17.02.03 CONDITION 2

Condition 2 can be implemented by the Officer-of-the-Deck, the First Officer or the Commanding Officer, and is automatically in effect when the ship goes to Red Alert.

Third Shift joins First and Second Shifts. An additional Security person will join each roving patrol. The number of roving patrols is doubled. Sentries at the Auxiliary Bridge and Brig will be augmented by another Security person. Two sentries are assigned to the Main Bridge, all Engineering

Spaces, all Transporter Rooms, Hangar Bay, Cargo Bay, Air Locks, and all Weapon Compartments.

When embarked, the MIMAD will be deployed as directed by the Detachment Commanding Officer.

All Security personnel will set phasers on the maximum NON-KILL level. The use of deadly force IS authorized by the senior person on the scene.

17.02.04 CONDITION 1

Condition 1 can only be implemented by the First Officer or the Commanding Officer. This condition is also known by the term "Intruder Alert."

All Security Interlocks throughout the ship are activated, preventing any unauthorized movement between decks or among corridors.

Security personnel will assume command and control authority over all personnel within the isolated spaces where they may find themselves. Immediate steps are to be taken to positively identify all personnel in their area of control through the use of palmprint, voice, and/or retina scan procedures.

When embarked, the MIMAD will be deployed as directed by the Detachment Commanding Officer.

Security personnel are to set phasers on KILL. The use of deadly force is automatically authorized during a Security Alert Condition 1 unless over-ridden by the officer implementing the condition.

Condition 1 can be rescinded only by the First Officer (if originated by that officer) or the Commanding Officer (whether originated by that officer or not).

18 LANDING PARTIES

There are six basic types of landing parties which may be called away in the course of the ship's mission. Although the following guidelines will apply to the majority of instances where a landing party is required, the Commanding Officer or officer-in-charge of the landing party may modify the composition and equipment carried as necessary to respond to specific circumstances.

The six types of landing parties, the reasons for their deployment, as well as their composition and equipment, are as follows.

18.01 DIPLOMATIC

When it becomes necessary for the ship's Commanding Officer to conduct diplomatic relations as a Federation representative, a special diplomatic landing party will be assembled.

The composition of this landing party will be at the Commanding Officer's complete discretion, except that one of the members must be a Diplomatic Services Specialist. In the case of a possible first contact mission, one of the members must be a First Contact Specialist.

Unless otherwise directed by the Commanding Officer, no weapons are to be carried by any member of a diplomatic landing party. Communicators will be carried by each member.

18.02 MEDICAL

Should a situation arise on a planet's surface where injured personnel are involved, a medical landing party, headed by either the Chief Medical Officer or a designated representative, and composed of seven additional personnel (five chosen from the Medical Department based on expected or foreseen specialty needs, and two Security Officers), will be dispatched as dictated by the Commanding Officer.

All members of this landing party will carry a communicator, and all medical personnel will carry a medikit. Security Officers will be armed with a hand phaser.

18.03 RESEARCH

The research landing party, assembled when intensive, on-site research is required, will be composed of individuals as assigned by the Chief Science Officer in consultation with the four division officers. Depending upon the situation, the research landing party may be transported to the appropriate location or utilize a shuttlecraft.

At least two security personnel and one Medical Department representative will be included in each landing party. Security personnel will be armed with hand phasers, and every member will carry a communicator. Medical personnel will have a medikit.

18.04 SEARCH AND RESCUE

In the event that an individual from a landing party or a complete landing party is missing, a search and rescue (SAR) landing party will, at the Commanding Officer's discretion, be called away to search for, locate, and return the missing individual(s).

The officer-in-charge of this party will be the Chief Security Officer. The remainder of the party is composed of five Security officers (assigned by the Chief Security Officer), and two Emergency Medical Technicians. Each member of this landing party will carry a communicator, and the Emergency Medical Technicians will each carry a medikit.

18.05 SECURITY

There may be occasions where a military objective must be secured on a planet or vessel. In this event, the Commanding Officer may, at his discretion, notify the MIMAD Commanding Officer to effect a security landing party, the composition and equipment carried by this landing party to be the prerogative of the MIMAD Commanding Officer.

In no case will the MIMAD Commanding Officer exceed the specific directions issued by the ship's Commanding Officer for the conduct and constraints placed upon this type of landing party.

18.06 SURVEY

Although the OMNI-Synse sensor package installed on the GALILEO increases the effective sensor range by 125% and improves resolution capabilities by 150% over the standard sensor suite, there is still no substitute for inserting a survey landing party onto a planet. The members of this type of party can utilize their tricorders for a close-up investigation and, at the same time, enter personal observations and conclusions into their Personal Logs.

Each member of the landing party is responsible, under the provisions of the Prime Directive, to avoid unnecessary contact with indigenous intelligent life forms and, failing this (should the landing party be discovered) to avoid displaying superior technology.   Each member of a survey landing party will carry a hand phaser, a communicator and a tricorder. Medical personnel will have a medikit in their possession .

The standard composition of a survey landing party is: one officer-in-charge, three scientific personnel, two Security personnel, and two Medical personnel.

19 VISITOR POLICY

While in orbit around a planet or docked in a starbase, there will be occasions when an individual not a part of the ship's company or Marine Detachment will desire to come aboard the ship for a visit.   Certain restrictions and procedures apply to all visitors.

19.01 ELIGIBILITY

The following personnel are eligible to visit the ship on a temporary basis:

> family members of assigned crew;

> citizens of the United Federation of Planets;

citizens of allied worlds not members of the United Federation of Planets; and,

members of Starfleet or other subsidiaries of the United Federation of Planets.

19.02 PROCEDURES

An initial request for visitation will be forwarded to the Chief Administrative Officer of the ship. After consultation with the Chief Security Officer, the Chief Administrative Officer of the ship will issue permission to the individual and arrange for quarters should the length of the visit indicate quarters will be necessary.

The Chief Administrative Officer will provide a listing of all visitors, their hosts (if any), quarters assignment (if any), and projected length of visit, to the Officer-of-the-Deck; the First Officer; the Chief Security Officer; the Commanding Officer, MIMAD (if embarked), the Chief Transporter Officer, and the Chief Flight Deck Control Officer. Each officer so notified will ensure that the visitor schedule is provided to every individual within their department who may have a need to know.

This visitor schedule will also be provided to each Security Officer. Any special considerations which may be appropriate, depending upon the status of the visitor (high-ranking Starfleet personnel or other civilian VIPs), will be included in the visitor schedule.

19.03 RESTRICTIONS

If requested, tours of the ship will be arranged by the ship's Orientation Officer in consultation with the Chief Security Officer. Unless special dispensation is granted by the Commanding Officer, the First Officer, the Officer-of-the-Deck, or the appropriate department head, no visitor will be allowed into these areas: Main Bridge, Auxiliary Bridge, Brig, any computer compartment, any weapons storage area, or any Science Department compartment designated by the Chief Science Officer as sensitive.

No visitor may be admitted to the Main Bridge or any Engineering Department propulsion space without an armed escort (either from the Security Department or the MIMAD [if embarked], depending upon the honors due such a visitor).

20 TRANSPORTATION POLICY

A limited number of quarters are available for transportation of personnel to destinations which do not interfere with the ship's mission or commitments. As with visitors, certain restrictions and procedures apply to personnel transportation.

20.01 ELIGIBILITY

The following personnel, listed in priority order, may be transported aboard the ship:

Members of the Federation Council and accompanying families;

Members of Starfleet and accompanying families under official change-of-duty-station orders;

Governmental officials and accompanying families of UFP-member worlds;

Members of Starfleet and accompanying families not under official change-of-duty-station orders;

Governmental officials and accompanying families of non-UFP member worlds;

Private citizens of the United Federation of Planets; and,

Individuals who are not citizens of the United Federation of Planets

20.02 PROCEDURES

An initial request for transportation will be forwarded to the Chief Administrative Officer of the ship. After consultation with the Chief Security Officer, the Chief Administrative Officer of the ship will issue permission to the individual and arrange for quarters assignment.

The Chief Administrative Officer will provide a listing of all visitors, their hosts (if any), quarters assignment, port of embarkation and debarkation, and projected length of visit, to the Officer-of-the-Deck; the First Officer; the Chief Security Officer; the Commanding Officer, MIMAD (if embarked); the Chief Transporter Officer and the Chief Flight Deck Control Officer. Each officer so notified will ensure that the manifest is provided to every individual within their department who may have a need to know.

This manifest will also be provided to each Security Officer. Any special considerations which may be appropriate, depending upon the status of the visitor (high-ranking Starfleet personnel or other civilian VIPs), will be included on the manifest.

20.03 RESTRICTIONS

If requested, tours of the ship will be arranged by the ship's Orientation Officer in consultation with the Chief Security Officer. Unless special dispensation is granted by the Commanding Officer, the First Officer, the Officer-of-the-Deck or the appropriate department head, no visitor will be allowed into these areas: Main Bridge, Auxiliary Bridge, Brig, any computer compartment, any weapons storage area, or any Science Department compartment designated by the Chief Science Officer as sensitive.

No visitor may be admitted to the Main Bridge or any Engineering Department propulsion space without an armed escort (either from Security Department or the MIMAD [if embarked], depending upon the honors due such a visitor).

21 Intra-Ship Communications

Intercoms are located in every compartment in the ship, in every station console, and at regular intervals in each corridor and access trunk. Each of these stations has its own, unique "address," by which the specific location from which an individual is speaking can be identified. Contacting an individual or specific station is as simple as speaking the individual's name into the audio pick-up section of a communication panel. The communications computer identifies the person for which the communication is intended by a specific code contained in memory.

A narrow-band signal is transmitted to a computer chip (transmitter/receiver) which is an integral part of every Starfleet personnel's ID card (which must be carried at all times), which, in turn, alerts the individual of the incoming communication. When so alerted, the individual needs only to go to an intercom panel and reply.

That specific panel is identified and the two communication panels, the originator and recipient, are connected via the Bridge (Master) circuit utilizing the appropriate secondary circuit based upon the originator's and the recipient's location. When both parties finish the conversation, the circuits are cleared for further use. If an individual's location is known, the originator can address the intercom panel at that location instead.

The communication circuit recognizes either a compartment number (such as 6P-16-2-T) or the common name of the location (such as Transporter Room 6P). All communications can be overridden or monitored by the Communications Station on the Bridge. The circuits and their dedicated areas are listed below.

Circuit	Area
1MC	Main Bridge
2MC	Main Reactor Compartment
3MC	Main Engineering
4MC	Secondary Engineering
5MC	Hangar Bay
6MC	Cargo Decks
7MC	Intercraft (All Units)
8MC	Damage Control
9MC	Fire Control
10MC	Transporter Rooms
11MC	Airlock/Hatchways
12MC	Sickbay
13MC	Auxiliary Bridge
14MC	Computer Rooms
15MC	Quarters (ALPHA Watch)
16MC	Quarters (BETA Watch)
17MC	Quarters (GAMMA Watch)
18MC	Security
19MC	MIMAD (when embarked)
20MC	Presently Unused

Intra-ship Communication Circuits

22 COMPARTMENT NUMBERING SYSTEM

In order to identify a specific compartment within the ship, a compartment designation system has been developed. Each compartment is assigned a unique code based on its location. There are four digits in this Compartment Designation Symbol (CDS) which are affixed to a plate secured to the door(s), hatch, and/or bulkhead of the compartment: Deck Number, Closest Frame Number, Closest Stanchion Number, and Usage Code Letter. When a compartment extends through one or more decks (the Flight Deck and the cargo holds are two examples), the lower Deck Letter is used.

Decks are the internal horizontal partitions of a ship, analogous to its "floors," and serve to divide it into a series of vertical levels. Each deck is assigned a number and a suffix indicating its location ("P" for the primary hull, "D" for the dorsal, and "S" for the secondary hull), beginning with Deck 1P for the Main Bridge, and continuing down to the lowest deck on the ship in sequence.

Frames are the main structural members of the ship which align, anchor, and strengthen the hull from within, supporting it much as the skeleton of a humanoid supports the muscles and skin. The Frame Number refers to the frame nearest the most forward (toward the bow of the ship) bulkhead of the compartment. GALILEO-class Starcruisers have a primary hull which is called an "interrupted-primary hull," and the frame numbering is slightly different than for a standard "disk-shaped" primary hull.

Frame Number 1 is located on the centerline at the bow of the ship. Frame Number 2 is located 18 degrees to starboard of that, frame Number 3 is 18 degrees from Number 2, and so forth until frame 7, which is the last frame which composes the circular shape. Aft of that is the extended hull (where the scientific laboratories are located) and the frames (now traverse rather than circular) extending toward the stern are numbered from 8 to 14. On the port side, just forward of the extended hull, the circular frames are numbered beginning with 15 and continue on to Frame 20 (which is 18 degrees to port of frame Number 1).

The secondary hull frame numbers begin with 21 and continue to increase by one until the hull is completed. Frames within the dorsal have the same numbers as the corresponding frames in the secondary hull.

Stanchions are the secondary structural members of the ship, which support and anchor all internal partitions and structures. Stanchion Numbers are from 1 to 10, and indicate how far the outboard bulkhead of the compartment is from the centerline (or central axis of the hull)—odd numbers for port side and even numbers for starboard side.

Usage Code Letter (UCL) indicates the general use of that compartment and are listed below.

UCL	Type	Description
A	Stowage	Lockers & Raw Materials Bins
B	Cargo	Holds & Container Spaces
C	Control	Bridge & Fire Control Operations
D	Computer	Computer & Electronics Spaces
E	Engineering	Engineering and Propulsion
F	Fuel-1	Anti-matter Containment
G	Fuel-2	Impulse Fuel Tankage
H	Hangar	Parking & Landing Bays
I	Fuel-3	Auxiliary Craft Fuel Stores
J	Presently Unused	Presently Unused
K	Hazard	Phaser Banks & Torpedo Bays
L	Living	Quarters & Recreation
M	Medical	Sickbay Facilities
N	Presently Unused	Presently Unused
O	Presently Unused	Presently Unused

UCL	Type	Description
P	Security	Armory, Offices & Brig
Q	Miscellaneous	Fabricators & Jeffries Tubes
R	Presently Unused	Presently Unused
S	Science	Research Laboratories
T	Transporter	Transporter Facilities
U	Presently Unused	Presently Unused
V	Presently Unused	Presently Unused
W	Water	Potable Water Tankage
X	Access	Corridors & Turboshafts
Y	Presently Unused	Presently Unused
Z	Presently Unused	Presently Unused

Compartment Usage Code Letters

23 MISCELLANEOUS SHIPBOARD GUIDELINES

This section contains information of importance to crewmembers which is not readily categorized into another part of this SORM.

23.01 PROTOCOL

An officer in Starfleet serves among other officers and over enlisted personnel. Some officers are superiors, some are equals, some subordinates. Likewise all enlisted personnel are subordinates. There are different forms and ways of interacting with each, depending upon relative seniority. However, something in common with all interactions is courtesy. Treating fellow personnel with proper respect and appropriate obedience is as essential and necessary a part of being an officer as is the giving of orders.

23.01.01   ADDRESSING OFFICERS

An officer is a peer among other officers. However, some formalities do pertain. When addressing or greeting a superior officer, he or she is addressed by rank and last name or simply by rank ("Captain Clayton" or "Captain"). An officer of equal rank is addressed by rank, rank and last name, or if the individual has given permission, by first name ("Lieutenant," "Lieutenant Kylye," or "Jon"). A subordinate officer is addressed by rank, rank and last name, the title "Mister," the title "Mister" and last name, or, if the individual has given permission, by first name ("Ensign," "Ensign Peters," "Mister," "Mister Peters," or "Kevin").

23.01.02   ADDRESSING ENLISTED PERSONNEL

An officer may use the following forms of address when speaking to an enlisted personnel. A Petty Officer is addressed by the rating, or rating and last name ("Petty Officer" or "Petty Officer Williams"). Chief Petty Officers are referred to by the title "Chief," or "Chief" and last name ("Chief," "Chief Brown"). A Senior Chief Petty Officer is referred to by the title "Senior Chief," or "Senior Chief" and name ("Senior Chief," "Senior Chief Brown"). A Master Chief Petty Officer is referred to by the title "Master Chief" or "Master Chief" and last name ("Master Chief," "Master Chief Burke").

23.01.03   ANSWERING A SUPERIOR OFFICER

When a superior officer gives an order, the correct response by the subordinate is "Aye, aye, sir" (the term "sir" is used for male and female officers alike). The superior may respond "Very well," or "Very Good," but not the subordinate. Saying "Aye, aye, sir" means three things:

(1) I heard the order;

(2) I understood the order; and,

(3) I will carry it out to the best of my ability.

To a question requiring a more verbose response, any answer must be respectful and must end with "sir." Similarly, an enlisted person will reply "Aye, aye, Chief" to a Chief Petty Officer.

23.02 GROOMING

Appearance and conduct should always reflect credit upon the individual, the ship, Starfleet, and the United Federation of Planets. The uniform should be of high quality, clean, and worn properly. Uniforms provided by Starfleet Quartermasters are regulation. If an uniform is made or purchased elsewhere, it is the wearer's responsibility to ensure it is regulation in pattern, appearance, fit, and quality. An adequate supply of clean uniforms must always be available. It is up to the wearer to ensure that uniforms are processed. If replicator-processed uniforms are worn, these concerns will be automatically met.

Unless authorized to wear civilian clothing, the complete uniform prescribed must be worn at all times. No articles such as pens, jewelry, combs, or similar items may be worn or carried exposed upon the uniform. Chronometers, inconspicuous rings, and conservative earrings are permitted. No extremes of dress are permitted.

23.03 UNIFORMS

The Uniform-of-the-Day will be announced to enlisted personnel by their petty officer prior to the commencement of each duty shift. Officers are expected to choose their own uniform using their knowledge of the watch's expected activities and the Commanding Officer's wishes. This may entail Class As (if, for example, an Admiralty inspection is scheduled), Class Bs (for a regular duty shift which may require some physical work), or Class Cs (depending upon any special hazards which may be expected).

23.03.01 CLASS A: (Dress)

Suitable for ceremonial functions and/or regular duties where prudent.

23.03.02 CLASS B: (Duty)

Suitable for regular duties and/or recreation and other off-duty activities.

23.03.03 CLASS C: (Special)

Worn when necessary in order to protect the wearer against special on-the-job hazards or for other requirements.

23.03.04 CLASS D: (Supplement)

Worn with the Uniform-of-the-Day solely at the discretion of the wearer. Examples of items which are considered Class D includes sweaters, wind breakers, and moisture-proof jackets.

23.04 DEPARTMENTAL COLORS

The following colors are assigned to the departments aboard ship, to be worn with the uniform as indicated by Starfleet regulations:

Department	Prescribed Color
Command	White
Operations	Gold
Engineering	Kelly Green
Science	Royal Blue
Medical	Purple
Security	Red

24 LEXICON

Abaft: Behind or farther aft; astern or toward the stern.

Abandon Ship: An order to all personnel to leave a stricken ship according to a pre-established plan of action, including lifeboats and auxiliary craft.

Abeam: At right angles to the centerline of and outside a ship.

Aboard: On or in a ship.

Above All, Knowledge: Motto of USS FERDINAND MAGELLAN (NCC-8892).

Adrift: Loose from mooring without power; out of place; out of control or lost.

Aft: In, near, or toward the stern of a ship.

After: That which is farthest aft.

Ahead: Forward of the bow of a ship.

Alongside: By the side of a ship.

Amidships: Indefinite area midway between bow and stern; in the middle portion of the ship, along the line of the keel.

Armament: Ship's weapons.

Ashore: On the beach or shore; not on the ship.

Astern: Toward the stern; an object or vessel that is abaft another vessel or object; directly behind a ship—Bearing 180 Mark 0.

A Tall Ship and a Star: Motto of USS CAROLUS LINNAEUS (NCC-8897).

Athwart: Across; at right angles to.

Authorized Grade: The minimum level (either officer or enlisted) of expertise which is appropriate for that specific position listed in the SMD.

Auxiliary: Extra or secondary.

Auxiliary Bridge: Located deep within the primary hull, also known as the Secondary Bridge. It replicates each station (with the exception of Science Station #2) on the Main Bridge, although not the physical layout. Its purpose is to provide the ship with command, control, and communication functions should the Main Bridge become damaged or uninhabitable for whatever reason. It is manned during Red Alerts and when so ordered by the Commanding Officer or Officer-of-the-Deck by qualified personnel as indicated in the SMD.

AVENGER, USS (NCC-1860): Heavy Frigate class ship (22 hulls). Refitted from starships of the SURYA (NCC-1850) class. They were designed primarily for use in various areas of the frontier, but have spent a majority of their active lives patrolling inner Federation territory. They are not equipped with extensive scientific or exploratory facilities.

Aye Aye: Reply to a command meaning "I understand and will comply."

Azimuth: See bearing.

Back: To go backwards.

Bank: Pair (or larger group) of similar weapons, fired as a unit.

Beam: Width; breadth; greatest athwartships width of a vessel; the extreme breadth of a ship.

Bearing: The direction of an object from an observer; first measured in degrees clockwise from dead ahead along the ship's X-Y plane. Second measured in degrees upward or downward from said X-Y plane.

Belay: To countermand or cancel a previous order or action.

BELKNAP, USS (NCC-2501): Strike Cruiser class ship (28 hulls). Replacements for the large number of perimeter action ships of the KIAGA and AGILIS classes (built during the Four Years War).

Below: Downward from a position.

Berth: Mooring space assigned to a ship.

Billet: Place or duty to which one is assigned.

Billet Sequence Code (BSC): An unique number assigned to each position listed in the SMD. Every individual assigned to a Starcruiser has a BSC which not only identifies that person's position but is also used by the Main Library Computer for security access to classified information.

Board: The act of going aboard a ship; a group of persons meeting for a specific purpose.

Bow: The forward end of a vessel.

Break Out: To unstow or prepare for use.

Bridge: Area in the superstructure from which the ship is controlled.

Brig: Prison on a ship, shore base, or starbase.

Broadside To: At right angles to the fore-and-aft line of a ship.

BSC: See Billet Sequence Code.

Bulkhead: A vertical partition (wall).

Buoy: A free-floating object in space designed for a specific task, such as communications, navigation, etc.

Burdened Vessel: That vessel which does not have the right of way.

c: A representation of the speed of light in a vacuum; 256,000 kilometers per second.

Cabin: Living quarters aboard a ship.

Captain's Gig: A small, easily stowed, five-passenger shuttle carried aboard larger starships which is used to ferry officers from ship to ship or from ship to station as a diplomatic gesture.

Carry Away: To break loose or tear loose.

Cathedral Unit: A test crew of personnel from the Starfleet Corps of Engineers and Starfleet Tactical which takes prototypes out on their shake-down cruise and evaluates the performance of all systems. Its name is taken from the secluded and cordoned-off sector reserved for this purpose.

Centerline: Imaginary line running from the ship's bow to its stern.

Chief Engineer: The senior member of the Engineering Department. This individual is responsible for the operation, maintenance, and repair of propulsion, weapons, and other operational systems aboard the ship.

Chief of Security: The senior member of the Security Department. This individual is responsible for protecting crew members on board the ship from potential injury or death, crewmembers off the ship on authorized landing parties, and to protect vital and sensitive areas of the ship or other facilities as designated by the Commanding Officer.

Chief Medical Officer: The senior member of the Medical Department. This individual is responsible for the health and welfare of every member of the crew, maintenance of all medical records, and other duties as assigned by the Commanding Officer.

Chain of Command: Succession of officers through which command is exercised from superior to subordinate. Aboard Starcruisers the chain of command is: Commanding Officer, First Officer, Operations Officer, Chief Science Officer, Chief Engineer, Chief Security Officer, Logistics Officer, Chief Medical Officer.

Chart: Usually holographic, a map of a volume of space showing stellar positions, traffic lanes, buoys, and navigational hazards.

Close Aboard: Near a ship.

Cold Iron: A condition of a vessel when it is not generating its own power; required power is provided by a starbase of another vessel.

Colors: United Federation of Planets' flag; command to raise or lower same.

Commanding Officer: The senior Starfleet officer aboard any commissioned vessel. A commanding officer has numerous responsibilities for both the ship and the area of the Federation in which the ship is operating. Among these are: military commander of any space sector in which the ship is located (including assembling ships in a fleet, mounting or reacting to an attack, and declaring the specific section to be in a state of combat); military governor of any colonies, outposts, or expeditions in the sector (final arbiter of the disbursement of materials, persons, or facilities to these outposts; authorized to settle any disputes over claims or territorial rights); supervisor of interstellar commerce, trade, or shipping in the section; aiding or restricting such trade according to the current conditions or require-ments of the Federation; the official representative of the Federation to any aligned or unaligned worlds or peoples in the section (having the power to open negotiations, arrange for trade and cultural exchange, and make temporary treaties of peace or partnership); responsible for the protection and stewardship of underdeveloped worlds and peoples according to the Prime Directive; serves as chief legal official of the Federation; possesses complete authority over all personnel assigned to his or her ship.

Commission: To activate a ship; a written order giving an officer his/her rank and authority.

Communicator: Personal transceiver.

Companionway: Deck opening giving access to a ladder.

Compartment: Interior space of a ship (room).

Conn: To direct a helmsman as to movement of helm; the act of controlling a ship.

CONSTELLATION, USS (NCC-1017): Starcruiser class ship (1 hull). Reported destroyed on shakedown cruise in 2299.

Convoy: A number of merchant ships sailing under the escort of warships; the act of escorting such ships.

Courage Leads to the Stars: Motto of USS NICOLAUS COPERNICUS (NCC-8896).

Course: A ship's desired direction and path of travel, not to be confused with heading.

Damage Control Locker: Compartments throughout the ship which have pre-positioned repair materials for use by DC Teams.

Damage Control Team: A group of specially trained individuals charged with the responsibility to provide a quick-reaction repair response to damage incurred during combat operations. Each DC Team is responsible for a specific area of the ship and operates out of a Damage Control Locker.

Damage Control Central: Exercises primary control over all damage control activities throughout the ship. Staffed by specially trained personnel and located in the primary hull.

Damage Control Secondary: Exercises secondary control over all damage control activities throughout the ship in support or (if necessary) as a replacement for DC Central. Staffed by specially trained personnel and located in the secondary hull.

DC Central: See Damage Control Central.

DC Secondary: See Damage Control Secondary.

DC Team: See Damage Control Team.

Dead Ahead: Directly ahead of a ship: Bearing 000 Mark 0.

Dead in the Water: Said of an underway ship that is making neither headway nor sternway, a hold-over expression from nautical ships.

Deck: Horizontal partition in a ship (floor).

Department: Personnel assigned to a starship are segregated into sub-organizations composed of crewmembers who have similar duties. For example, all personnel who operate and maintain the propulsion units are assigned to Engineering Department.

Department Head: The senior officer assigned to a department aboard ship. This individual is charged with ensuring the personnel assigned to that specific department execute their duties in an efficient and effective manner.

Derelict: Abandoned vessel.

Displacement: The volume of space, expressed in metric tons, which a star ship occupies. It is one facet of describing the physical parameters of a vessel. The term originally referred to the amount of fluid displaced by a vessel operating in a water environment but has been adapted to star ships.

Division: Within a department, specialists who are assigned responsibility for a specific area are divided into divisions. For example in the Engineering Department, those specialists who operate and maintain the offensive and defensive weapons systems of the ship are in the Weapons Systems Division of the Engineering Department.

Division Officer: The senior officer assigned to a division within a department aboard ship. This individual reports directly to the Department Head and is responsible for ensuring the personnel assigned to that specific division execute their duties in an efficient and effective manner.

Dock: Any structure which serves as a mooring point for a vessel; the act of so mooring.

Draft: The extreme height of a vessel.

Dreadnought: Before the introduction of the Starcruiser class exploration and scientific research vessel, it was the largest and best-armed class of vessel in Starfleet. It is normally used to provide overwhelming offensive capabilities to a Task Force engaged in hostile actions.

Drift: The deviation of a ship from its plotted course or position.

Drydock: A dock which is equipped to refit or rebuild ships.

Durolithium: One of the newly developed materials of which the hull of the GALILEO-class Starcruiser is constructed.

ECHO (**E**nhanced **C**ollimination via **H**armonic **O**scillation): An enhancement to the main offensive weapons of Starcruisers which increases the destructive force delivered to the target. Activation of this ECHO system must not be effected when the phaser bank is set on "stun," since it is lethal to all known life forms inhabiting the target area.

Embark: To go on board a ship preparatory to sailing.

End on: Head-to-head or stem-to-stem.

Engineering Officer of the Watch: The senior Engineer during each of the three duty shifts. This individual assumes all duties and responsibilities of the Chief Engineer while on watch.

ENTERPRISE-II: Heavy Cruiser class ship (14 hulls). Originally classified as Exploratory Cruisers, designation changed to Heavy Cruiser in 2287. Almost identical to the earlier ENTERPRISE class.

EOOW: See Engineering Officer of the Watch.

Eugene's Limit: The theoretical upper limit of faster-than-light travel. At Eugene's Limit, the power requirement to maintain a warp field approaches infinity, and is, therefore, unattainable.

Fantail: The after end of a Hangar Deck.

First Fleet: A numbered fleet under Starfleet Command responsible for a specific sector (Quadrant 1) of Federation territory. Various ships and personnel are assigned to First Fleet in order to carry out this mission.

First Officer: Second-in-command aboard a commissioned Starfleet vessel. In the absence or incapacitation of the Commanding Officer, assumes all responsibilities and authority vested in that position. The equivalent officer on a shore station or Starbase is called an Executive Officer.

Flag Officer: An officer above the rank of Captain, so called because he/she is entitled to fly his/her personal flag which, by stars, indicates rank.

Flash Team: A team put together by a responsible officer composed of individuals possessing specialized knowledge, abilities, or characteristics. Organized for a specific purpose, a Flash Team has limited existence and is disbanded after the particular task has been accomplished.

Flat: Grating or partial deck to provide walking and working surfaces, used extensively in engineering spaces.

Fleet: An organization of ships and structures (such as starbases and/or drydocks), all under one command.

Forward: Toward the bow; opposite of aft.

Fourth Fleet: A numbered fleet under Starfleet Command responsible for a specific sector (Quadrant 4) of Federation territory. Various ships and personnel are assigned to Fourth Fleet in order to carry out this mission.

Frame: The secondary structural members of a ship's hull.

Gravitic Mine: Developed by the Romulan Star Empire for use against the Federation, they are scattered throughout Federation space at unknown locations. Composed of a matter/anti-matter warhead and a proximity fuse, they are detonated by the changes in gravity caused by the passage of a medium-sized or larger starship. Because the detonation of one often causes a chain-reaction, exploding all gravitic mines in the area, they are extremely hazardous; however, they are often unreliable and fail to detonate.

GRISSOM, USS (NCC-638): GRISSOM-class scientific research vessels were the mainstay of the scientific research and exploration fleet of Starfleet until age and new technologies rendered it obsolete. A small vessel, it carried a crew of approximately 80, including scientists and crewmembers.

Hangar: Space used for landing, launching, and parking auxiliary craft.

Hard Over: Extreme turn to one side.

Hatch: Removable cover to a gangway.

Head: Compartment containing sanitary facilities.

Heading: The direction a ship is facing while underway.

Headway: Forward motion of a ship.

Hold: Large cargo space aboard a ship.

Home Fleet: An unnumbered fleet under Starfleet Command responsible for a specific sector (Quadrant 0) of Federation territory. Various ships and personnel are assigned to the Home Fleet in order to carry out this mission.

Humbrolite: One of the newly developed materials of which the hull of the GALILEO-class Starcruiser is constructed.

Impulse Drive: Secondary means of propulsion for Starfleet vessels. It is utilized for sub-light travel in close proximity to planets and other bodies. Maximum speed which can be attained in impulse drive is less than the speed of light.

Inboard: Toward the centerline.

Jury Rig: Any makeshift device or repair.

Keel: The main strength members of a ship, upon which the frames and hull plates depend.

Keelhaul: To reprimand severely.

Knock Off: To cease what is being done; to stop work.

Launch: To maneuver a vessel from dock or planet surface.

Lie To: Said of a vessel when underway with no way on.

Log: Official record in which data or events that occurred during a watch are stored; a personal log is a record made by an individual ancillary to an official log.

Mega-phaser: An improved version of the standard phaser emplacements utilized as the main weapon system of a star ship. It produces a larger destructive force than standard phasers and is often called a "cannon." Mega-phasers aboard GALILEI-class Starcruisers are further improved by being ECHO enhanced.

Memory Alpha: A planetoid set up by the Federation as a central library containing the total cultural history and scientific knowledge of all Federation members.

MEMT: See Mobile Emergency Medical Team.

Mobile Emergency Medical Team: Composed of a physician, a nurse, and other medically trained personnel; assigned to various locations throughout the ship during Red Alerts to respond to medical emergencies within a specified area of responsibility.

Moor: To anchor or make fast to a dock.

Motto: A word, phrase, or sentence chosen as expressive of the goals or ideals of an organization or individual. Most vessels, starbases, and shore stations in Starfleet have a motto. See elsewhere for a listing of all GALILEO-class Starcruiser mottos.

Mychromium: One of the newly developed materials of which the hull of the GALILEO-class Starcruiser is constructed.

Nadir: Directly below: Bearing 000 Mark -90.

Neutral Zones: Zones in space set up by treaties in which neither belligerent has control or is allowed to enter. In many cases, violation of a neutral zone is paramount to declaring war.

Officer-of-the-Deck: The senior officer of the watch on the Main Bridge of a starship. This officer has full responsibility for the safe operation, course, defense, and any other matter delegated by the Commanding Officer in the absence of the Commanding Officer or First Officer from the Main Bridge.

Ordnance: General term meaning weapons systems.

Outboard: Away from the centerline; toward the side of the vessel, or outside the vessel entirely.

Overhaul: To overtake another ship on the same heading; to undertake extensive repairs or modifications to a ship or facility.

Overhead: On a ship, equivalent to the ceiling of a building ashore; ships have overheads rather than ceilings.

Overticite: One of the newly developed materials of which the hull of the GALILEO-class Starcruiser is constructed.

Padd (Portable Archival Digital Device): A data storage device used by personnel to record information when the more powerful tricorder is inappropriate; i.e., personal logs. Both vocal and stylus input are available. Data transfer between the Main Library Computer and individual padds can be effected. There are five models of padds, classified by memory capacity: Model 1 (1GB), Model 2 (2GB), Model 3 (3GB), Model 4 (4GB) and Model 5 (5GB).

Parsec (**Par**allax of one **sec**ond of arc): A measure of distance in deep space, a parsec is equivalent to 3.26 light years.

Passageway: Corridor or hallway on a ship.

Pass the Word: To repeat an order or information to all hands.

Paumerium: One of the newly developed materials of which the hull of the GALILEO is constructed.

Phaser (PHASed Energy Rectification): The main energy weapon of Starfleet star ships; the main personal weapon utilized for offense and defense by individuals in Starfleet. They are usually issued for Landing Party use in an unknown situation or when circumstances dictate.

Photon Torpedo: First developed by the Romulans, these projectiles with a matter/anti-matter warhead deliver a more concentrated and destructive payload on a target than ship's phasers.

Pipe Down: An order to keep silent.

Pipe the Side: Ceremony in which sideboys are drawn up and the boat-swain's pipe is blown when a high-ranking officer or distinguished visitor comes aboard.

Pitch: Rotation of a ship on the Y axis.

Plank Owner: A person who served aboard a ship at its commissioning.

Polyhedrium: One of the newly developed materials of which the hull of the GALILEO-class Starcruiser is constructed.

Port: To the left of the centerline when facing forward, Bearing 270 Mark 0; a major spaceship landing field or orbital station.

Prima inter Pares: Ancient Earth Latin for "First Among Her Equals." The motto of USS GALILEO GALILEI (NCC-8888).

Primary Hull: The "main hull" of a starship, where the majority of tactical and strategic control compartments are located. In Starfleet, the predominate shape of this primary hull is a disk, or saucer. In emergencies, it can be detached from the secondary hull and, using the impulse drive units, travel at limited speeds, landing on a planet's surface to provide living accommodations for the crew until assistance arrives. Once landed, the primary hull can be retrieved only by specialized equipment found in shipbuilding or conversion facilities.

Prime Directive: Starfleet General Order Number 1. Directs all Starfleet personnel to obey a policy of non-interference in the affairs of any culture. It is the guiding principle in relationships between Federation and non-Federation races.

Quarters: Living spaces aboard ship.

Range: Distance between observer and object.

Red Alert: Also called Battle Stations or General Quarters. The most extreme level of readiness for a ship in response to imminent danger on the ship, in the immediate vicinity, or possible offensive or defense action which might result in personnel casualties or damage to the ship.

Relieve: To take the place (duties) of another.

Restricted Maneuverability Zone (RMZ): The use of warp drive within a planetary system causes an interaction between the ship's warp field envelope and the star's hydrostatic equilibrium. This, in turn, impacts on the star's Rayleigh Scattering Ratio. As a result, ships are restricted to the use of impulse power within the system's RMZ. This RMZ is different for each star system and is determined by the star's class, number of planets within the system and their relative positions, and other, more technical measurements.

Rig: To set up any piece of equipment.

Roll: Rotation of a ship on the X axis.

Running Lights: Navigational lights mounted on a ship's exterior; red for port, green for starboard, white at other, specific points; standard flashing rate for white running lights is 1.1 seconds.

SCIEX MK-I MOD-0: Main Bridge module for GALILEO-class Starcruisers which is designed to support the scientific research and exploratory mission of the Starcruiser class vessel. Included in this module are two dedicated science stations which directly control the OMNI-Synse sensor package.

Scuttle: A round, air-tight hatch; to self-destruct a vessel.

Secondary Hull: The second hull of a Starfleet starship in which the majority of the Engineering spaces and Warp Drive propulsion units are located. A Hangar Bay, cargo bays, and storage facilities are also usually located in the secondary hull.

Second Fleet: A numbered fleet under Starfleet Command responsible for a specific sector (Quadrant 2) of Federation territory. Various ships and personnel are assigned to Second Fleet in order to carry out this mission.

Secondary Sickbay: An auxiliary medical facility located in the secondary hull to handle local casualties or non-critical patient overflow from Main Sickbay.

Second Star on the Right and On Till Morning: Motto of USS LEONARDO DA VINCI (NCC-8894).

Sensor: General term for a device which gathers data either in a passive or active mode.

Shakedown: The training of a new crew; the trial run of a prototype, or a vessel having finished refit.

Shipshape: Neat; clean; taut.

Ship's Manning Document: A document which lists each position aboard a ship by the Billet Sequence Code along with the authorized grade for each and the name and rank of the individual filling that position.

Ship's Organization and Regulations Manual (SORM): The document which details the organization of a ship and outlines the regulations to be followed while a member of the crew.

Sickbay: Ship's medical spaces and facilities.

Side Boy: Member of the honor guard welcoming VIPs.

Skylark: To engage in irresponsible behavior.

SMD: See Ship's Manning Document.

SORM: See Ship's Organization and Regulations Manual.

Squadron: A fighting unit composed of up to 20 vessels having a common mission or capabilities.

Stanchions: Girders used to support bulkheads, attached to frames.

Stand By: Prepare for.

Starboard: To the right of centerline when facing forward, Bearing 090 Mark 0.

Starcruiser: Newest class of Starfleet vessel, designed for scientific research and exploration of deep space. The original Starcruiser vessel was USS CONSTELLATION (NCC-1017), reported lost with all hands on its shakedown cruiser in 2299.

Stateroom: A living compartment for an officer or passenger.

Station: A crew member's place of duty; position of a ship in a formation.

Stern: The aftermost part of a ship.

Stow: To pack articles of cargo in a space or container.

Supercargo: A individual who is physically on a ship but who is not part of the ship's company; a passenger. The term is usually reserved for members of Starfleet who are being transported on the ship, but it can be applied to any individual.

Task Force: Temporary grouping of units under one commander; formed for the purpose of carrying out a specific operation or mission.

There Are Always Alternatives: Motto of USS PAUL A. M. DIRAC (NCC-8898).

Thermocoat: A non-reflective, non-refractive coating applied to the outer hull of Starfleet vessels under 90,000 metric tons displacement. Designed to aid the ship in evading detection by other sensor systems; application of this covering becomes cost- and weight-prohibitive, as well as ineffective, for large vessels.

The Universe Is Relative: Motto of USS ALBERT EINSTEIN (NCC-8889).

Third Fleet: A numbered fleet under Starfleet Command responsible for a specific sector (Quadrant 3) of Federation territory. Various ships and personnel are assigned to Third Fleet in order to carry out this mission.

Today the Improbable, Tomorrow the Impossible: Motto of USS STEPHEN W. HAWKING (NCC-8899).

To Know the Infinite Unknown: Motto of USS JACQUES-IVES COUSTEAU (NCC-8893).

To Sail the Farthest Reaches: Motto of USS CAPTAIN JAMES COOK (NCC-8890).

To Search, Therefore to Find: Motto of USS CHANDRASEKHAR (NCC-8891).

Tractor Beam: A repulsor/attractor device on starships which can manipulate large objects at a distance by altering their inertia as desired.

Traverse Frame: Structural member which extends outward from the centerline of a vessel.

Turn In: Retire to quarters; return articles to storage; report an individual for violation of a rule or regulation.

Turn Out: Get out of bed; order out a work party.

Turn To: Start work.

Underway: A ship is underway when not moored or in planetary orbit. She need not be actually moving; she is underway so long as she lies free in space and under her own power.

Veer: To swerve suddenly from a previous heading.

Void: An empty compartment or container; deep space.

Warp Drive: The main method of propulsion for Starfleet vessels which uses an intricate mix of matter and antimatter to attain speeds in excess of c. Warp drive is utilized for travel only outside of star systems because of the effect the drive can have on planets (see Restricted Maneuverability Zone).

Warp Factor: A measurement of speed in faster-than-light travel using the Warp Drive.

Watch: One of the three eight-hour periods into which a day is divided on board ships and some orbital facilities; watches are denoted by the terms ALPHA, BETA and GAMMA.

We Can Because We Think We Can: Motto of USS ZEFRAM COCHRANE (NCC-8895).

Wormhole: An anomaly in space created by any one of several unique combinations of circumstances. A ship caught by a wormhole is seldom able to escape its effects.

Yaw: Rotation of a ship on the Z axis.

Yellow Alert: A condition of heightened readiness to respond to potentially dangerous situations either on the ship or in the immediate vicinity.

Zenith: Directly above, Bearing 000 Mark +90.

25 SHIP MANNING DOCUMENT

This section lists every authorized position on a GALILEO-class Starcruiser. In some cases, one individual may fill more than one Billet, depending upon secondary skills, experience, or specialized training. Therefore, the SMD does not necessarily reflect the number of personnel actually assigned to a particular ship.

The Billet Sequence Code (BSC) is a six-digit number assigned each Billet (or authorized position). The first two digits indicate the department, the second two digits indicate a division within the department, and the last two digits indicate the individual's sequential billet number.

A Billet Title is assigned to each authorized billet. This title describes in general terms what that particular job entails. For a more detailed explanation of each Billet Title, refer to the appropriate sections elsewhere in this SORM.

Each BSC has an Authorized Grade assigned, which indicates the minimum experience level expressed as a rank for each position. Because of variable Starfleet manning levels, it is not always possible to fill every billet at the Authorized Grade. The one-up/one-down concept of ship manning allows the Billet to be filled by an individual who may be one rank higher or one rank lower than the Authorized Grade. For example, if the Authorized Grade for a particular BSC calls for a Lieutenant (junior grade), or LTJG, either a Lieutenant (LT), Lieutenant (junior grade), or an Ensign (ENS) may be assigned to that BSC.

The only exception to this policy is BSC 000001 (Commanding Officer) and BSC 000002 (First Officer). Personnel assigned to these two positions will always be a Captain and a Commander, respectively.

Information on Shifts can be found elsewhere in this SORM.

During a Red Alert, all personnel are assigned a specific location or membership on a specific "team." If the Red Alert assignment is a team, then the individual filling that BSC must have received appropriate training and have maintained the necessary qualifications.

Damage Control Teams are stationed at strategic locations throughout the ship. These DC Teams make initial repairs to battle damage suffered by the ship during combat operations.

Mobile Emergency Medical Teams (MEMT) are composed of individuals with primary or secondary medical training. Each MEMT is stationed at a designated Battle Dressing Station and is responsible for a specific area of the ship. A MEMT provides First Aid and triage services to injured personnel within their respective areas.

Security Teams provide protection and security services to various critical areas throughout the ship. Each member of a team has a specific duty within their team's assigned perimeters.

BSC	Billet Title	Authorized Grade	Shift	Red Alert Assignment
	COMMAND DEPARTMENT			
000001+	Commanding Officer	CAPT	N	Main Bridge
000002+	First Officer	CDR	N	Auxiliary Bridge
	Special Assistants Division			
000101	Orientation Officer	LT	S	DC Team
000102	Religious Assistance Specialist	LT	A	Main Sickbay
000103	Religious Assistance Specialist	LTJG	B	Secondary Sickbay
000104	Religious Assistance Specialist	LTJG	G	Chapel
000105	Diplomatic Services Specialist	LCDR	N	DC Team
000106	Diplomatic Services Specialist	LT	N	DC Team
000107	First Contact Specialist	LCDR	N	DC Team
000108	First Contact Specialist	LT	N	DC Team
000109	Ship's Historian/Archivist	LT	S	Admin Office
000110	Educational Services Specialist	LTJG	A	DC Team
000111	Educational Services Specialist	LTJG	B	DC Team
000112	Educational Services Specialist	LTJG	G	DC Team
	Administration Division			
000201	Chief Administrative Officer	LT	S	Admin Office
000202	Chief Yeoman	MCPO	N	DC Central
000203	Senior Enlisted Advisor	MCPO	N	Auxiliary Bridge
000204	Captain's Yeoman	SCPO	N	Main Bridge
000205	Operations Department Yeoman	PO1	N	OPS Center
000206	Shuttle Systems Section Yeoman	PO3	N	Shuttlebay
000207	Engineering Department Yeoman	PO1	N	Main Engineering
000208	Science Department Yeoman	PO1	N	Science Office
000209	Life Sciences Division Yeoman	PO2	N	DC Central

BSC	Billet Title	Authorized Grade	Shift	Red Alert Assignment
000210	Physical Sciences Division Yeoman	PO2	N	DC Team
000211	Social Sciences Division Yeoman	PO2	N	DC Team
000212	Space Sciences Division Yeoman	PO2	N	DC Team
000213	Security Department Yeoman	PO2	N	Security Office
	OPERATIONS DEPARTMENT			
010001+	Operations Officer	CDR	N	OPS Center
010002+	Assistant Operations Officer	LCDR	N	OPS Center
	Shuttle Operations Division			
010101	Chief Shuttlecraft Support Officer	LCDR	N	Shuttlebay
010102	Shuttlecraft Support Officer	LT	S	Shuttlebay
010103	Shuttlecraft Pilot	LTJG	A	Shuttlecraft
010104	Shuttlecraft Pilot	LTJG	B	Shuttlecraft
010105	Shuttlecraft Pilot	LTJG	G	Shuttlecraft
010106	Shuttlecraft Pilot	ENS	A	Shuttlecraft
010107	Shuttlecraft Pilot	ENS	B	Shuttlecraft
010108	Shuttlecraft Pilot	ENS	G	Shuttlecraft
010109	Chief Shuttlecraft Mechanic	LTJG	S	Shuttlebay
010110	Shuttlecraft Mechanic	CPO	A	M&R Bay
010111	Shuttlecraft Mechanic	CPO	B	M&R Bay
010112	Shuttlecraft Mechanic	CPO	G	M&R Bay
010113	Shuttlecraft Mechanic	PO2	A	M&R Bay
010114	Shuttlecraft Mechanic	PO2	B	M&R Bay
010115	Shuttlecraft Mechanic	PO2	G	M&R Bay
010116	Shuttlecraft Support Technician	PO3	A	M&R Bay
010117	Shuttlecraft Support Technician	PO3	B	M&R Bay
010118	Shuttlecraft Support Technician	PO3	G	M&R Bay

BSC	Billet Title	Authorized Grade	Shift	Red Alert Assignment
	Navigation Division			
010201+	Chief Navigator	LCDR	N	Main Bridge
010202+	Assistant Navigator	LT	A	Auxiliary Bridge
010203+	Navigator	LTJG	B	DC Central
010204+	Navigator	ENS	G	DC Secondary
010205	Navigational Systems Analyst	ENS	A	DC Central
010206	Navigational Systems Analyst	ENS	B	DC Secondary
010207	Navigational Systems Analyst	ENS	G	DC Secondary
010208	Navigational Systems Technician	LTJG	A	DC Central
010209	Navigational Systems Technician	ENS	B	DC Secondary
010210	Navigational Systems Technician	ENS	G	DC Secondary
	Helm Division			
010301+	Chief Helmsman	LCDR	N	Main Bridge
010302+	Assistant Helmsman	LT	A	Auxiliary Bridge
010303+	Helmsman	LTJG	B	DC Secondary
010304+	Helmsman	LTJG	G	DC Secondary
010305	Helm Systems Analyst	LTJG	A	DC Central
010306	Helm Systems Analyst	ENS	B	DC Secondary
010307	Helm Systems Analyst	ENS	G	DC Secondary
010308	Helm Systems Technician	LTJG	A	DC Central
010309	Helm Systems Technician	ENS	B	DC Secondary
010310	Helm Systems Technician	ENS	G	DC Secondary
	Communications Division			
010401+	Chief Communications Officer	LCDR	A	Main Bridge
010402+	Assistant Communications Officer	LT	B	Auxiliary Bridge
010403	Computer Systems Analyst	LT	S	Comm Center

BSC	Billet Title	Authorized Grade	Shift	Red Alert Assignment
010404+	Communications Systems Officer	LTJG	G	Comm Center
010405	Subspace Equipment Specialist	ENS	A	Comm Center
010406	Subspace Equipment Specialist	ENS	B	Comm Center
010407	Subspace Equipment Specialist	ENS	G	Comm Center
010408	Internal Equipment Specialist	ENS	A	Comm Center
010409	Internal Equipment Specialist	ENS	B	Comm Center
010410	Internal Equipment Specialist	ENS	G	Comm Center
010411	Equipment Technician	PO1	A	Comm Center
010412	Equipment Technician	PO1	B	Comm Center
010413	Equipment Technician	PO1	G	Comm Center
Flight Deck Division				
010501	Chief Flight Deck Control Officer	LCDR	N	Flight Deck Control
010502	Flight Deck Control Specialist	LTJG	A	Flight Deck Control
010503	Flight Deck Control Specialist	LTJG	B	Flight Deck Control
010504	Flight Deck Control Specialist	LTJG	G	Flight Deck Control
010505	Equipment/Supply Technician	ENS	A	Shuttlebay
010506	Equipment/Supply Technician	ENS	B	M&R Bay
010507	Equipment/Supply Technician	ENS	G	M&R Bay
010508	Maintenance/Repair Specialist	PO1	A	Shuttlebay
010509	Maintenance/Repair Specialist	PO1	B	M&R Bay
010510	Maintenance/Repair Specialist	PO1	G	M&R Bay
ENGINEERING DEPARTMENT				
020001	Chief Engineer	CDR	N	Main Engineering
020002	Assistant Chief Engineer	LCDR	N	Main Bridge
020003	Computer Systems Analyst	LT	S	Main Engineering

BSC	Billet Title	Authorized Grade	Shift	Red Alert Assignment
	Engineering Systems Division			
020101	Chief Propulsion Systems Officer	LCDR	N	Main Engineering
020102	Warp Drive Engineer	LT	A	Main Engineering
020103	Warp Drive Engineer	LT	B	Main Engineering
020104	Warp Drive Engineer	LT	G	DC Team
020105	Impulse Drive Engineer	LTJG	A	Main Engineering
020106	Impulse Drive Engineer	LTJG	B	Main Engineering
020107	Impulse Drive Engineer	LTJG	G	DC Team
020108	Thruster Unit Specialist	ENS	A	Main Engineering
020109	Thruster Unit Specialist	ENS	B	Main Engineering
020110	Thruster Unit Specialist	ENS	G	DC Team
020111	Dilithium Crystals Specialist	LTJG	A	Main Engineering
020112	Dilithium Crystals Specialist	LTJG	B	Main Engineering
020113	Dilithium Crystals Specialist	LTJG	G	DC Team
020114	Matter/Anti-matter Specialist	ENS	A	Main Engineering
020115	Matter/Anti-matter Specialist	ENS	B	Main Engineering
020116	Matter/Anti-matter Specialist	ENS	G	DC Party
020117	Maintenance Technician	CPO	A	Main Engineering
020118	Maintenance Technician	CPO	B	Main Engineering
020119	Maintenance Technician	CPO	G	Main Engineering
	Transporter Systems Division			
020201	Chief, Transporter Systems Officer	LCDR	N	Main Engineering
020202	Transporter Systems Officer	LT	S	Main Engineering
020203	Transporter Systems Specialist	CPO	A	Transporter #1
020204	Transporter Systems Specialist	CPO	B	Transporter #2
020205	Transporter Systems Specialist	CPO	G	Transporter #3

BSC	Billet Title	Authorized Grade	Shift	Red Alert Assignment
020206	Transporter Systems Specialist	PO1	A	Transporter #4
020207	Transporter Systems Specialist	PO1	B	Transporter #5
020208	Transporter Systems Specialist	PO1	G	Transporter #6
020209	Maintenance Technician	PO2	A	Transporter #1
020210	Maintenance Technician	PO2	B	Transporter #2
020211	Maintenance Technician	PO2	G	Transporter #3
020212	Maintenance Technician	PO3	A	Transporter #4
020213	Maintenance Technician	PO3	B	Transporter #5
020214	Maintenance Technician	PO3	G	Transporter #6
	Operational Systems Division			
020301+	Chief, Operational Systems Officer	LCDR	N	Main Bridge
020302	Operational Systems Officer	LT	S	Auxiliary Bridge
020303	Structural Engineering Specialist	LTJG	S	DC Team
020304	Acoustical Systems Technician	ENS	S	DC Team
020305	Life-Support Systems Technician	ENS	A	Main Engineering
020306	Life-Support Systems Technician	ENS	B	DC Team
020307	Life-Support Systems Technician	ENS	G	DC Team
020308	Environmental Systems Technician	ENS	A	Main Engineering
020309	Environmental Systems Technician	ENS	B	DC Team
020310	Environmental Systems Technician	ENS	G	DC Team
020311	Bio-Systems Specialist	LTJG	S	DC Team
020312	Deflector Shields Specialist	LTJG	A	Main Engineering
020313	Deflector Shields Specialist	ENS	B	DC Team
020314	Deflector Shields Specialist	ENS	G	DC Team
020315	Tractor Beam Specialist	LTJG	A	Main Engineering
020316	Tractor Beam Specialist	ENS	B	DC Team

BSC	Billet Title	Authorized Grade	Shift	Red Alert Assignment
020317	Tractor Beam Specialist	ENS	G	DC Team
020318	Auxiliary Systems Specialist	LT	A	Main Engineering
020319	Auxiliary Systems Specialist	LTJG	B	DC Team
020320	Auxiliary Systems Specialist	ENS	G	DC Team
020321	Sensor Systems Specialist	LTJG	A	Main Engineering
020322	Sensor Systems Specialist	LTJG	B	DC Team
020323	Sensor Systems Specialist	LTJG	G	DC Team
	Weapons Systems Division			
020401+	Chief, Weapons Systems Officer	LCDR	N	Main Bridge
020402	Phaser Systems Engineer	LT	A	Auxiliary Bridge
020403	Phaser Systems Specialist	LTJG	B	Weapon Room 1
020404	Phaser Systems Specialist	ENS	G	Weapon Room 2
020405	Photon Torpedo Specialist	LTJG	A	Weapon Room 1
020406	Photon Torpedo Specialist	LTJG	B	Weapon Room 2
020407	Photon Torpedo Specialist	LTJG	G	Magazine
020408	Maintenance Technician	CPO	A	Weapon Room 1
020409	Maintenance Technician	CPO	B	Weapon Room 2
020410	Maintenance Technician	CPO	G	Magazine
020411	Maintenance Technician	PO1	A	Weapon Room 1
020412	Maintenance Technician	PO2	B	Weapon Room 2
020413	Maintenance Technician	PO3	G	Magazine
	Closed-Systems Recovery Division			
020501	Chief, Closed-Systems Officer	LT	N	DC Central
020502	Refurbisher	PO1	A	DC Team
020503	Refurbisher	PO1	B	DC Team

BSC	Billet Title	Authorized Grade	Shift	Red Alert Assignment
020504	Refurbisher	PO1	G	DC Team
020505	Janitorial Services	PO2	A	DC Team
020506	Janitorial Services	PO2	B	DC Team
020507	Janitorial Services	PO2	G	DC Team
020508	Sanitation Maintenance Technician	PO1	A	DC Team
020509	Sanitation Maintenance Technician	PO2	B	DC Team
020510	Sanitation Maintenance Technician	PO3	G	DC Team
020511	Hydroponics Specialist	CPO	A	Botanical Gardens
020512	Hydroponics Specialist	PO1	B	Botanical Gardens
020513	Hydroponics Specialist	PO2	G	Botanical Gardens
020514	Chief, Fabrications Officer	LT	S	DC Central
020515	Fabrication Specialist	LTJG	A	DC Team
020516	Fabrication Specialist	LTJG	B	DC Team
020517	Fabrication Specialist	LTJG	G	DC Team
020518	Fabrication Specialist	ENS	A	DC Team
020519	Fabrication Specialist	ENS	B	DC Team
020520	Fabrication Specialist	ENS	G	DC Team
	SCIENCE DEPARTMENT			
030001+	Chief Science Officer	CDR	N	Main Bridge
030002+	Assistant Science Officer	LCDR	S	Main Bridge
	Life Sciences Division			
030101+	Chief Life Sciences Officer	LCDR	N	Auxiliary Bridge
030102	Computer Systems Analyst	LT	S	DC Central
030103	Agronomist	ENS	S	DC Team
030104	Anatomist	ENS	S	DC Team
030105	Biochemist	ENS	S	DC Team

BSC	Billet Title	Authorized Grade	Shift	Red Alert Assignment
030106	Bioengineer	ENS	S	DC Team
030107	Biologist	ENS	S	DC Team
030108	Biotechnologist	ENS	S	DC Team
030109	Botanist	ENS	S	DC Team
030110	Cryogenics Officer	LTJG	S	Main Sickbay
030111	Ecologist	ENS	S	DC Team
030112	Embryologist	ENS	S	DC Team
030113	Entomologist	ENS	S	DC Team
030114	Geneticist	ENS	S	DC Team
030115	Herpetologist	ENS	S	DC Team
030116	Horticulturalist	ENS	S	DC Team
030117	Ichthyologist	ENS	S	DC Team
030118	Limnologist	ENS	S	DC Team
030119	Mastologist	ENS	S	DC Team
030120	Marine Biologist	ENS	S	DC Team
030121	Microbiologist	ENS	S	DC Team
030122	Ornithologist	ENS	S	DC Team
030123	Paleontologist	ENS	S	DC Team
030124	Pathologist	ENS	S	DC Team
030125	Physiologist	ENS	S	DC Team
030126	Toxicologist	ENS	S	DC Team
030127	Zoologist	ENS	S	DC Team
	Physical Sciences Division			
030201+	Chief Physical Sciences Officer	LCDR	N	Auxiliary Bridge
030202+	Chief Computer Systems Analyst	LT	S	DC Central
030203	Computer Systems Analyst	LTJG	S	DC Team

BSC	Billet Title	Authorized Grade	Shift	Red Alert Assignment
030204	Analytical Chemist	ENS	S	DC Team
030205	Cartographer	ENS	S	DC Team
030206	Chemical Oceanographer	ENS	S	DC Team
030207	Climatologist	ENS	S	DC Team
030208	Economic Geologist	ENS	S	DC Team
030209	Exploration Physicist	ENS	S	DC Team
030210	Geochemist	ENS	S	DC Team
030211	Geochronologist	ENS	S	DC Team
030212	Geodesist	ENS	S	DC Team
030213	Geological Oceanographer	ENS	S	DC Team
030214	Geomagnetician	ENS	S	DC Team
030215	Geomorphologist	ENS	S	DC Team
030216	Hydrologist	ENS	S	DC Team
030217	Inorganic Chemist	ENS	S	DC Team
030218	Mineralogist	ENS	S	DC Team
030219	Organic Chemist	ENS	S	DC Team
030220	Paleomagnetician	ENS	S	DC Team
030221	Paleontologist	ENS	S	DC Team
030222	Palynologist	ENS	S	DC Team
030223	Physical Meteorologist	ENS	S	DC Team
030224	Physical Oceanographer	ENS	S	DC Team
030225	Planetologist	ENS	S	DC Team
030226	Seismologist	ENS	S	DC Team
030227	Soil Scientist	ENS	S	DC Team
030228	Stratigrapher	ENS	S	DC Team
030229	Synoptic Meteorologist	ENS	S	DC Team

BSC	Billet Title	Authorized Grade	Shift	Red Alert Assignment
030230	Volcanologist	ENS	S	DC Team
	Social Sciences Division			
030301+	Chief Social Sciences Officer	LCDR	N	DC Team
030302	Computer Systems Analyst	LT	S	DC Team
030303	Anthropologist	LTJG	S	DC Team
030304	Anthropologist	ENS	S	DC Team
030305	Archeologist	ENS	S	DC Team
030306	Archeologist	LTJG	S	DC Team
030307	Curator	ENS	S	DC Team
030308	Demographic Researcher	ENS	S	DC Team
030309	Ethnoscientist	ENS	S	DC Team
030310	Geographer	ENS	S	DC Team
030311	Geographer	LTJG	S	DC Team
030312	Historian	ENS	S	DC Team
030313	Legal Specialist	ENS	S	DC Team
030314	Paleoanthropologist	ENS	S	DC Team
030315	Political Scientist	ENS	S	DC Team
030316	Sociologist	ENS	S	DC Team
	Space Sciences Division			
030401+	Chief Space Sciences Officer	LCDR	N	DC Team
030402	Computer Systems Analyst	LT	S	DC Team
030403	Aeronautical Specialist	ENS	S	DC Team
030404	Aerospace Specialist	ENS	S	DC Team
030405	Astronautical Specialist	ENS	S	DC Team
030406	Astronomer	LTJG	S	DC Team
030407	Astronomer	ENS	S	DC Team

BSC	Billet Title	Authorized Grade	Shift	Red Alert Assignment
030408	Astronomer	ENS	S	DC Team
030409	Astronomical Cartographer	ENS	S	DC Team
030410	Astronomical Cartographer	ENS	S	DC Team
030411	Astrophysicist	ENS	S	DC Team
030412	Cosmologist	ENS	S	DC Team
030413	Cosmologist	LTJG	S	DC Team
030414	Elementary Particle Physicist	ENS	S	DC Team
030415	Molecular Physicist	ENS	S	DC Team
030416	Nuclear Physicist	ENS	S	DC Team
030417	Plasma Physicist	ENS	S	DC Team
030418	Pulsar Specialist	ENS	S	DC Team
030419	Quasar Specialist	ENS	S	DC Team
030420	Thermodynamic Physicist	ENS	S	DC Team
	MEDICAL DEPARTMENT			
040001	Chief Medical Officer	CDR	N	Main Sickbay
040002	Assistant Medical Officer	LCDR	S	Secondary Sickbay
040003	Computer Systems Analyst	LT	S	Main Sickbay
040004	Archivist/Librarian	PO1	A	Main Sickbay
040005	Archivist/Librarian	PO2	B	Secondary Sickbay
040006	Archivist/Librarian	PO3	G	DC Team
040007	Physician	LT	A	Main Lounge
040008	Physician	LT	B	Main Sickbay
040009	Physician	LT	G	Main Sickbay
040010	Physician	LT	S	Secondary Sickbay
040011	Chief Nurse	LCDR	S	Main Sickbay
040012	Nurse	LT	A	MEMT 1

BSC	Billet Title	Authorized Grade	Shift	Red Alert Assignment
040013	Nurse	LT	B	Main Sickbay
040014	Nurse	LT	G	Main Lounge
040015	Nurse	LTJG	A	MEMT 2
040016	Nurse	LTJG	B	Main Sickbay
040017	Nurse	LTJG	G	Main Sickbay
040018	Nurse	ENS	A	MEMT 3
040019	Nurse	ENS	B	Secondary Sickbay
040020	Nurse	ENS	G	Secondary Sickbay
040021	Laboratory Technician	PO1	A	Main Sickbay
040022	Laboratory Technician	PO1	B	Main Sickbay
040023	Laboratory Technician	PO1	G	Secondary Sickbay
040024	Laboratory Technician	PO2	A	MEMT 1
040025	Laboratory Technician	PO2	B	MEMT 2
040026	Laboratory Technician	PO2	G	MEMT 3
040027	Maintenance & Repair Technician	PO1	A	Main Sickbay
040028	Maintenance & Repair Technician	PO1	B	Main Sickbay
040029	Maintenance & Repair Technician	PO1	G	Secondary Sickbay
040030	Maintenance & Repair Technician	PO2	A	MEMT 1
040031	Maintenance & Repair Technician	PO2	B	MEMT 2
040032	Maintenance & Repair Technician	PO2	G	MEMT 3
040033	Emergency Medical Technician	CPO	A	MEMT 1
040034	Emergency Medical Technician	CPO	B	MEMT 2
040035	Emergency Medical Technician	CPO	G	MEMT 3
040036	Emergency Medical Technician	PO1	A	MEMT 1
040037	Emergency Medical Technician	PO1	B	MEMT 2
040038	Emergency Medical Technician	PO1	G	MEMT 3

BSC	Billet Title	Authorized Grade	Shift	Red Alert Assignment
	SECURITY DEPARTMENT			
050001	Chief Security Officer	CDR	N	Security Office
050002	Assistant Security Officer	LCDR	S	Security Officer
	Security Personnel Division			
050101	Security Watch Officer	LT	A	Security Team 1
050102	Security Officer	ENS	A	Security Team 1
050103	Security Officer	ENS	A	Security Team 1
050104	Security Officer	ENS	A	Security Team 1
050105	Security Officer	ENS	A	Security Team 1
050106	Security Officer	ENS	A	Security Team 1
050107	Security Officer	ENS	A	Security Team 1
050108	Security Officer	ENS	A	Security Team 1
050109	Security Officer	ENS	A	Security Team 1
050110	Security Officer	ENS	A	Security Team 1
050111	Security Officer	ENS	A	Security Team 1
050112	Security Officer	ENS	A	Security Team 1
050113	Security Officer	ENS	A	Security Team 1
050114	Security Officer	ENS	A	Security Team 1
050115	Security Officer	ENS	A	Security Team 1
050116	Security Watch Officer	LT	B	Security Team 2
050117	Security Officer	ENS	B	Security Team 2
050118	Security Officer	ENS	B	Security Team 2
050119	Security Officer	ENS	B	Security Team 2
050120	Security Officer	ENS	B	Security Team 2
050121	Security Officer	ENS	B	Security Team 2
050122	Security Officer	ENS	B	Security Team 2

BSC	Billet Title	Authorized Grade	Shift	Red Alert Assignment
050123	Security Officer	ENS	B	Security Team 2
050124	Security Officer	ENS	B	Security Team 2
050125	Security Officer	ENS	B	Security Team 2
050126	Security Officer	ENS	B	Security Team 2
050127	Security Officer	ENS	B	Security Team 2
050128	Security Officer	ENS	B	Security Team 2
050129	Security Officer	ENS	B	Security Team 2
050130	Security Officer	ENS	B	Security Team 2
050131	Security Watch Officer	LT	G	Security Team 3
050132	Security Officer	ENS	G	Security Team 3
050133	Security Officer	ENS	G	Security Team 3
050134	Security Officer	ENS	G	Security Team 3
050135	Security Officer	ENS	G	Security Team 3
050136	Security Officer	ENS	G	Security Team 3
050137	Security Officer	ENS	G	Security Team 3
050138	Security Officer	ENS	G	Security Team 3
050139	Security Officer	ENS	G	Security Team 3
050140	Security Officer	ENS	G	Security Team 3
050141	Security Officer	ENS	G	Security Team 3
050142	Security Officer	ENS	G	Security Team 3
050143	Security Officer	ENS	G	Security Team 3
050144	Security Officer	ENS	G	Security Team 3
050145	Security Officer	ENS	G	Security Team 3
	Personal Weapons Division			
050201	Chief Personal Weapons Officer	LT	S	Armory
050202	Assistant Weapons Officer	LTJG	N	Armory

BSC	Billet Title	Authorized Grade	Shift	Red Alert Assignment
050203	Armorer	ENS	A	Armory
050204	Armorer	ENS	B	Armory
050205	Armorer	ENS	G	Armory
050206	Phaser Repair Specialist	ENS	A	Armory
050207	Phaser Repair Specialist	ENS	B	Armory
050208	Phaser Repair Specialist	ENS	G	Armory
050209	Weapons Systems Instructor	LTJG	A	Security Team 1
050210	Weapons Systems Instructor	LTJG	B	Security Team 2
050211	Weapons Systems Instructor	LTJG	G	Security Team 3
	MARINE DETACHMENT			
050301	Commanding Officer	MAJ	N	As Assigned
050302	Assistant	SGTM	N	As Assigned
	First Platoon (Light Weapons)			
050303	Platoon Sergeant	SSGT	S	As Assigned
050304	First Section Leader	SGT	A	As Assigned
050305	1st Fire Team Leader	CPL	A	As Assigned
030506	Senior Light Weapons Specialist	LCPL	A	As Assigned
030507	2nd Fire Team Leader	CPL	A	As Assigned
030508	Light Weapons Specialist	PVT	A	As Assigned
030509	3rd Fire Team Leader	LCPL	A	As Assigned
030510	Light Weapons Specialist	PVT	A	As Assigned
030511	Second Section Leader	SGT	B	As Assigned
030512	1st Fire Team Leader	CPL	B	As Assigned
030513	Senior Light Weapons Specialist	LCPL	B	As Assigned
030514	2nd Fire Team Leader	CPL	B	As Assigned
030515	Light Weapons Specialist	PVT	B	As Assigned

BSC	Billet Title	Authorized Grade	Shift	Red Alert Assignment
030516	3rd Fire Team Leader	CPL	B	As Assigned
030517	Light Weapons Specialist	PVT	B	As Assigned
030518	Third Section Leader	SGT	G	As Assigned
030519	1st Fire Team Leader	CPL	G	As Assigned
030520	Senior Light Weapons Specialist	LCPL	G	As Assigned
030521	2nd Fire Team Leader	CPL	G	As Assigned
030522	Light Weapons Specialist	PVT	G	As Assigned
030523	3rd Fire Team Leader	CPL	G	As Assigned
030524	Light Weapons Specialist	PVT	G	As Assigned
	Second Platoon (Heavy Weapons)			
030525	Platoon Sergeant	SSGT	S	As Assigned
030526	First Section Leader	SGT	A	As Assigned
030527	1st Fire Team Leader	CPL	A	As Assigned
030528	Senior Heavy Weapons Specialist	LCPL	A	As Assigned
030529	2nd Fire Team Leader	CPL	A	As Assigned
030530	Heavy Weapons Specialist	PVT	A	As Assigned
030531	3rd Fire Team Leader	CPL	A	As Assigned
030532	Heavy Weapons Specialist	PVT	A	As Assigned
030533	Second Section Leader	SGT	B	As Assigned
030534	1st Fire Team Leader	CPL	B	As Assigned
030535	Senior Heavy Weapons Specialist	LCPL	B	As Assigned
030536	2nd Fire Team Leader	CPL	B	As Assigned
030537	Heavy Weapons Specialist	PVT	B	As Assigned
030538	3rd Fire Team Leader	CPL	B	As Assigned
030539	Heavy Weapons Specialist	PVT	B	As Assigned

BSC	Billet Title	Authorized Grade	Shift	Red Alert Assignment
030540	Third Section Leader	SGT	G	As Assigned
030541	1st Fire Team Leader	CPL	G	As Assigned
030542	Senior Heavy Weapons Specialist	LCPL	G	As Assigned
030543	2nd Fire Team Leader	CPL	G	As Assigned
030544	Heavy Weapons Specialist	PVT	G	As Assigned
030545	3rd Fire Team Leader	CPL	G	As Assigned
030546	Heavy Weapons Specialist	PVT	G	As Assigned
	Third Platoon (Reconnaissance)			
030547	Platoon Sergeant	SSGT	S	As Assigned
030548	First Section Leader	SGT	A	As Assigned
030549	1st Reconnaissance Team Leader	CPL	A	As Assigned
030550	Advanced Scout	LCPL	A	As Assigned
030551	2nd Reconnaissance Team Leader	CPL	A	As Assigned
030552	Scout	PVT	A	As Assigned
030553	3rd Reconnaissance Team Leader	CPL	A	As Assigned
030554	Scout	PVT	A	As Assigned
030555	Second Section Leader	SGT	B	As Assigned
030556	1st Reconnaissance Team Leader	CPL	B	As Assigned
030557	Advanced Scout	LCPL	B	As Assigned
030558	2nd Reconnaissance Team Leader	CPL	B	As Assigned
030559	Scout	PVT	B	As Assigned
030560	3rd Reconnaissance Team Leader	CPL	B	As Assigned
030561	Scout	PVT	B	As Assigned
030562	Third Section Leader	SGT	G	As Assigned
030563	1st Reconnaissance Team Leader	CPL	G	As Assigned
030564	Advanced Scout	LCPL	G	As Assigned

BSC	Billet Title	Authorized Grade	Shift	Red Alert Assignment
030565	2nd Reconnaissance Team Leader	CPL	G	As Assigned
030566	Scout	PVT	G	As Assigned
030567	3rd Reconnaissance Team Leader	CPL	G	As Assigned
030568	Scout	PVT	G	As Assigned
	LOGISTICS DEPARTMENT			
060001	Chief Logistics Officer	CDR	N	Logistics Office
060002	Assistant Logistics Officer	LCDR	S	Logistics Office
	Stores Division			
060101	Chief Quartermaster	LT	N	DC Central
060102	Storekeeper	CPO	A	DC Central
060103	Storekeeper	PO1	A	Emergency Stores
060104	Storekeeper	PO2	A	Emergency Stores
060105	Storekeeper	CPO	B	Emergency Stores
060106	Storekeeper	PO1	B	Emergency Stores
060107	Storekeeper	PO2	B	Emergency Stores
060108	Storekeeper	CPO	G	Emergency Stores
060109	Storekeeper	PO1	G	Emergency Stores
060110	Storekeeper	PO2	G	Emergency Stores
	Food Division			
060201	Nutritionist	LT	S	DC Team
060202	Replicator Maintenance	CPO	A	DC Team
060203	Replicator Maintenance	PO1	A	DC Team
060204	Replicator Maintenance	CP0	B	DC Team
060205	Replicator Maintenance	PO1	B	DC Team
060206	Replicator Maintenance	CPO	G	DC Team
060207	Replicator Maintenance	PO1	G	DC Team

BSC	Billet Title	Authorized Grade	Shift	Red Alert Assignment
060208	Culinary Specialist	CPO	A	DC Team
060209	Culinary Specialist	CPO	B	DC Team
060210	Culinary Specialist	CPO	G	DC Team
	Recreation Division			
060301	Recreation Officer	LT	S	Main Lounge
060302	Lounge Services Officer	LTJG	N	Main Lounge
060303	Lounge Services Personnel	PO1	A	Main Lounge
060304	Lounge Services Personnel	PO2	A	MEMT 1
060305	Lounge Services Personnel	PO3	A	MEMT 2
060306	Lounge Services Personnel	PO1	B	MEMT 3
060307	Lounge Services Personnel	PO2	B	MENT 1
060308	Lounge Services Personnel	PO3	B	MENT 2
060309	Lounge Services Personnel	PO1	G	MEMT 3
060310	Lounge Services Personnel	PO2	G	Secondary Sickbay
060311	Lounge Services Personnel	PO3	G	DC Team
060312	Athletics Officer	ENS	A	DC Team
060313	Athletics Officer	ENS	B	DC Team
060314	Athletics Officer	ENS	G	DC Team
060315	Personal Services Specialist	PO1	A	DC Team
060316	Personal Services Specialist	PO2	A	DC Team
060317	Personal Services Specialist	PO3	A	DC Team
060318	Personal Services Specialist	PO1	B	DC Team
060319	Personal Services Specialist	PO2	B	DC Team
060320	Personal Services Specialist	PO3	B	DC Team
060321	Personal Services Specialist	PO1	G	DC Team
060322	Personal Services Specialist	PO2	G	DC Team

BSC	Billet Title	Authorized Grade	Shift	Red Alert Assignment
060323	Personal Services Specialist	PO3	G	DC Team
060324	Individual Sports Specialist	PO2	A	DC Team
060325	Individual Sports Specialist	PO2	B	DC Team
060326	Individual Sports Specialist	PO2	G	DC Team
060327	Team Sports Specialist	PO2	A	DC Team
060328	Team Sports Specialist	PO2	B	DC Team
060329	Team Sports Specialist	PO2	G	DC Team
060330	Music Instructor	PO1	S	DC Team
060331	Art Instructor	PO1	S	DC Team
060332	Thespian Instructor	PO1	S	DC Team

26 INDEX

27 ACKNOWLEDGEMENTS

My thanks go to Major Ronald V. Murray, USMC (Retired) for developing the organization of the Mobile Independent Marine Action Detachment.

As usual, Eric Kristiansen has provided valuable advice and insight. His contributions were critical. I also wish to thank my wife for putting up with stuff spread throughout the house, and Eugenia P. Miller, PhD, for her careful continuity checks.

Illustrations on Pages 30, 32, 33, 35, 37, 38, 39, 42, 43, 45, 47, 61, 92, 93, 94, 95, 96, 97, 98 (with minor modifications) are reprinted with permission from Jackill's *Star Fleet Reference Manual*, Ships of the Fleet, Volume I, written and illustrated by Eric Kristiansen.

28 SHIPBOARD DATA SHEET

Name	
Rank	
Serial Number	
Billet Sequence Code	
Billet Title	
Security Clearance	
Security Password	
Division Assigned	
Division Officer	
Department Assigned	
Department Head	
Shift	
Yellow Alert Assignment	
Red Alert Assignment	
Abandon Ship Assignment	
Quarters Assigned	
Additional Notes	